Prompt Engineering for Everyone

A Self-Taught, Human-Centered Approach to AI Programming

Hamid Tavakoli

Apress®

Prompt Engineering for Everyone: A Self-Taught, Human-Centered Approach to AI Programming

Hamid Tavakoli
AI Fairness CIO,
London, UK

ISBN-13 (pbk): 979-8-8688-2337-4 ISBN-13 (electronic): 979-8-8688-2338-1
https://doi.org/10.1007/979-8-8688-2338-1

Managing Director, Apress Media LLC: Welmoed Spahr
Acquisitions Editor: Shivangi Ramachandran
Project Manager: Jessica Vakili

Distributed to the book trade worldwide by Springer Science+Business Media New York, 1 New York Plaza, New York, NY 10004. Phone 1-800-SPRINGER, fax (201) 348-4505, e-mail orders-ny@springer-sbm.com, or visit www.springeronline.com. Apress Media, LLC is a Delaware LLC and the sole member (owner) is Springer Science + Business Media Finance Inc (SSBM Finance Inc). SSBM Finance Inc is a **Delaware** corporation.

For information on translations, please e-mail booktranslations@springernature.com; for reprint, paperback, or audio rights, please e-mail bookpermissions@springernature.com.

Apress titles may be purchased in bulk for academic, corporate, or promotional use. eBook versions and licenses are also available for most titles. For more information, reference our Print and eBook Bulk Sales web page at http://www.apress.com/bulk-sales.

If disposing of this product, please recycle the paper

For everyone who has been excluded because they do not speak the language of the dominant group.

And for those who work to ensure every voice is heard, every story matters, and every system becomes more just.

This book is for you.

Table of Contents

About the Author

Hamid Tavakoli is a research and development specialist with over two decades of experience in applied AI, optics, and ophthalmic medical devices. A qualified optometrist and optical design engineer, he holds postgraduate qualifications in applied AI and linguistics from leading UK universities. As Managing Director of Optics AI Ltd, he leads ethical, human-centered AI innovation and regularly shares insights on prompt engineering across academic and professional platforms worldwide.

About the Technical Reviewer

Faria Hossain is a fully funded PhD researcher in Virtual Production at London South Bank University (LSBU), UK, currently in the thesis-writing stage. Her research focuses on applied computer vision, artificial intelligence, volumetric video compression, and quality-of-experience (QoE) modelling, with particular emphasis on dataset creation, perceptual evaluation, and reproducible research methodologies for immersive media systems.

She holds an MSc in Data Science (Distinction) from London South Bank University, where she achieved the highest dissertation score, as well as MSc and BSc degrees in Computer Science and Engineering from Daffodil International University, Bangladesh. Her academic background spans machine learning, image and video processing, data analytics, networking, and software systems.

She has extensive teaching experience across Levels 4–7 in UK higher education. She currently works as an hourly paid lecturer at London South Bank University, teaching MSc-level modules including Artificial Intelligence Foundations and Calculus, Computer Vision, Research Methods, and Statistical Analysis and Modelling and serving as module leader for an MSc Data Science module. She is also a part-time lecturer at David Game Higher Education, contributing to computing-related modules. Previously, from January 2025 to September 2025, she taught undergraduate computing modules at Regent College London, including Databases, Advanced Programming, Artificial Intelligence, and Computing

Infrastructure. She also has prior lecturing experience at Daffodil International University. Her teaching approach is research-led, inclusive, and closely aligned with industry practice.

Her research contributions include the creation of widely used, publicly available datasets published on IEEE DataPort, including the *Forgery Image Dataset* and the *VVC-Compressed Volumetric Video Dataset*. She has authored and co-authored multiple peer-reviewed publications in IEEE conferences and journals covering media forensics, dataset benchmarking, volumetric video, and applied computing systems. She also maintains open research code and experimental pipelines on GitHub to support transparency and reuse.

In addition to her academic work, she has industry experience as a data and systems analyst (consultant), contributing to the design of data architectures and analytics pipelines for an educational AR and digital game project using cloud-based SQL and NoSQL technologies. This industry engagement informs both her teaching and applied research.

Preface

The development of large language models has brought about a profound shift in how machines interpret and generate human language. What once appeared to be a statistical breakthrough, automating fluency, has evolved into a general-purpose infrastructure for communication, reasoning, and interaction across professions and institutions. At the center of this transformation lies the prompt: a linguistic artifact that not only initiates model output but also governs tone, framing, and the ethical shape of a response.

Prompt engineering has evolved in parallel with this shift. It began as a creative strategy, then matured into a reproducible technique, and now stands as an applied discipline requiring critical reflection and design responsibility. Today, prompts operate silently in the background of widely used systems, guiding chatbots, influencing summarization, shaping feedback tools, and mediating everyday decisions. Yet their content remains largely invisible. Their logic is often opaque, their assumptions unexamined, and their impact untraceable.

This book addresses that challenge. It treats prompt engineering not as a niche technical specialty but as a foundational form of interaction design, one that directly influences how information is structured, how decisions are framed, and how users are included or excluded. Grounded in systems thinking and human-centered practice, each chapter seeks to clarify the functional role of prompting while foregrounding its ethical consequences.

The book is written for learners across fields, students, educators, clinicians, policymakers, content professionals, and public service workers, many of whom are encountering these systems without formal technical training. No programming background is required. The goal is to support prompt literacy as a professional capacity: a set of transferable skills for working effectively and responsibly with AI in structured, high-stakes, or public-facing environments.

To that end, the book is designed as a self-guided learning resource. Concepts are introduced progressively, illustrated through real-world analogies, and reinforced through reflective exercises and examples. While no single definition of prompt engineering can capture a moving target, the chapters provide a principled foundation for evaluating the structure, purpose, and reliability of prompts.

Framing prompt engineering as a public literacy is a deliberate choice. It reflects the belief that the future of AI systems will be shaped not only by model architecture but by the human values encoded in their instructions. Prompts are no longer private commands; they are policy mechanisms, design levers, and cultural signals. They deserve to be written with intention, revised with care, and understood within the broader systems that govern trust and meaning.

A prompt, once deployed, becomes part of institutional memory. It determines what users see, how information is framed, and whose voices are made legible. These are not incidental outcomes; they are the terrain on which prompt engineering will either mature as a responsible discipline or reinforce systems of exclusion and opacity.

This book is one contribution to the development of that discipline. It is offered as a reference, a training aid, and a pedagogical companion to those shaping AI use in practice. The future of prompt engineering will not be determined by models alone; it will be shaped by the intent, judgment, and care of the people who write the prompts.

Wishing you an insightful journey and purposeful prompting.

Acknowledgments

This book could not have been written without the intellectual, professional, and emotional support of many individuals and communities. Though authored independently, it stands on the foundation of shared effort, open dialogue, and collective inquiry.

I am especially grateful to the educators, researchers, developers, and activists whose early efforts to humanize artificial intelligence provided the ethical groundwork for this work. Their dedication to fairness, access, and accountability helped shape not only the content but also the core values expressed throughout these pages.

To the global communities of learners, particularly those working across languages, disciplines, and institutional boundaries, your questions, feedback, and practical experiments have given this project its most urgent clarity. It is in dialogue with you that the idea of *prompt engineering for everyone* took root and gained its meaning.

I also acknowledge the vital contributions of public institutions, international organizations, and open-source collaborators whose research, tools, and frameworks for inclusive AI development informed many of the positions and examples presented across chapters.

A special and deeply personal thanks goes to my daughter, Saba. Without her encouragement, curiosity, and quiet support, this book may never have been completed. What began as a simple checklist to guide her learning in prompt engineering eventually grew into the foundation of this entire volume. Her insight and enthusiasm continue to animate the human-centered approach at the heart of this work.

Finally, I extend my deepest gratitude to everyone who supported this journey, not only through technical expertise and intellectual exchange but also through personal generosity, patience, and a belief in the value of this effort. The result is more than a practice guide; it is a statement of shared responsibility. Whatever holds lasting value in these pages belongs as much to the broader community as to the author.

Author's Note

This work did not begin as a technical manual. It started as a response to a quiet, persistent gap: many were being told to "learn to use AI," yet few were given the tools to understand what that meant, linguistically, ethically, or professionally. Too often, those outside the AI industry were excluded from shaping systems that were already shaping them.

I wrote this book for those individuals, for learners who must now interact with systems they did not build, in languages they did not choose, and in contexts where AI increasingly mediates decisions that affect lives.

Prompt engineering is not the exclusive domain of engineers. It is a design practice that increasingly belongs to teachers, clinicians, civic actors, translators, and advocates, those whose work centers not on optimization, but on care. Writing this book has been an act of that same care: an effort to create space for thoughtful, inclusive, and principled prompting in a world that urgently needs it.

I am deeply grateful to the readers who have accompanied this journey. May your work continue to shape systems that are fairer, clearer, and more human.

CHAPTER 1

Why Prompt Engineering Standards Matter

Cutting-edge language models like ChatGPT, Claude, Gemini, and LLaMA are revolutionizing multiple sectors spanning education, healthcare, legal practice, financial services, and routine digital interactions. These advanced systems can produce coherent academic papers, convert text between languages, develop software programs, and replicate specialist cognitive processes across varied disciplines. However, each sophisticated output fundamentally depends on a frequently overlooked component: the user's initial input instructions.

A prompt is not merely a user message. The prompt serves as both a structural framework and a communicative guide, shaping the model's objectives, stylistic approach, breadth of response, and operational boundaries. Much like architects working from detailed blueprints or surgeons adhering to strict procedural guidelines, effective LLM interactions require meticulously crafted prompts to generate meaningful, logically consistent results. Prompt design determines not only what the model says, but how well it aligns with human expectations.

H. Tavakoli, *Prompt Engineering for Everyone*,
https://doi.org/10.1007/979-8-8688-2338-1_1

Despite this central role, most prompts today are composed with little formal method. Vague, ambiguous, or underspecified instructions remain widespread, often leading to verbose or inaccurate responses. These inefficiencies result not only in wasted time and computational resources but also in real-world consequences, including user confusion, inflated costs, and the propagation of biased or misleading information. Academic and technical communities have advanced prompt engineering, but public, institutional, and educational settings have not yet adopted a shared professional standard.

This chapter introduces the rationale for such a standard. It frames prompt design not as a casual act of experimentation, but as a core literacy for working responsibly with advanced language models. The study argues that prompting now constitutes an essential component of digital fluency, alongside coding, information literacy, and critical reading, in all domains of learning.

To support this argument, the chapter outlines the structural, economic, environmental, and ethical consequences of operating without standards. It highlights how even minor improvements in prompt design can reduce computational costs, improve clarity and reuse, and minimize harm. More importantly, it positions prompt engineering as a human-centered, design-oriented practice with implications for equity, sustainability, and trust in AI systems.

This book develops the Prompt Anatomy Blueprint as a framework to support this vision. In doing so, it offers a path toward shared language, rigor, and care in how we engage with language models, not only as users but as responsible stewards of their impact.

1.1 The Hidden Cost of Poor Prompts

Every interaction with a large language model carries a cost, measured not only in time and attention but in tangible computational and financial resources. These models process text as tokens, the discrete units of language that represent words or subwords. Each token requires energy to process and compute, and at a global scale, even minor inefficiencies can become significant.

Consider the following example: a user types, *"Can you tell me a bit about AI in general? Thanks a lot, in advance."* Although this request may seem polite and open-ended, it offers little guidance. The model responds with a long, general-purpose answer that may or may not meet the user's actual intent. What could have been a summary turns into a broad essay. The user spends more time reading, the model spends more time generating, and both sides contribute to unnecessary computational load.

Now imagine this kind of vague prompt occurring not once, but billions of times per day. The resulting waste of energy, money, and time is staggering. Each prompt typically generates thousands of tokens, and each token consumes processing power. At scale, these interactions account for billions of tokens processed daily across cloud infrastructure. Even a slight refinement in prompt construction, reducing token volume by just 5% or 10%, could save millions of dollars annually, substantially lowering energy consumption.

Cloud infrastructure, electricity consumption, and the environmental footprint of large-scale model deployment directly determine these concrete costs. Inefficient prompts squander resources that could otherwise support more urgent applications in healthcare, education, scientific research, or public services, representing a tangible opportunity loss.

Prompting, then, is not just a stylistic concern. It is a form of resource management. A well-written prompt yields focused responses, minimizes waste, and contributes to a more sustainable deployment of language models. In contrast, careless prompting places a hidden burden on both technological systems and the societies that rely on them.

This book begins with efficiency, not because it is the only concern, but because it is a *foundational* one. Writing clearly and purposefully is the first ethical act of responsible AI interaction. Structured prompting is not only about extracting better answers but also about building systems that are leaner, fairer, and more aligned with the values of sustainable design. One recent analysis of large-scale language models (including OpenAI system reports and university studies such as UCL evaluations) confirms that clearer prompting can measurably reduce energy usage by reducing token generation and processing loads.

1.2 Proven Savings from Better Prompts

Structured prompt design not only improves the quality of model output but also produces measurable efficiencies. Practitioners across sectors demonstrate that even modest refinements in prompt construction can significantly reduce token usage, processing time, and computational strain.

Organizations working in business, education, and software development have reported tangible gains when adopting basic prompt design strategies. For instance, prompts that specify task structure, such as dividing an instruction into steps or defining the expected output format, consistently yield shorter and more relevant responses. In enterprise environments, teams have reduced token consumption by over 30% simply by rephrasing vague prompts into more targeted, purposeful instructions. These optimizations require no changes to the underlying models. They depend only on the clarity of human input.

In technical domains, the impact is even more dramatic. Structured prompting in code generation and data analysis workflows has reduced token usage by over 90% in some cases by eliminating unnecessary modifiers, compressing verbose language, and employing compact yet precise phrasing. When a prompt includes a clear role, audience, and task, rather than leaving these elements implicit, the model can operate more efficiently, avoiding generalizations or excessive elaboration.

These improvements do not require specialist expertise. Educators, researchers, and professionals outside of computer science have achieved comparable gains by applying common-sense adjustments, such as defining tone, indicating format, or adding examples when needed. What these cases share is a shift in mindset, from prompting as an improvisational task to prompting as an act of design.

At its core, this shift transforms prompting into a form of optimization. The goal extends beyond eliciting better responses to achieving them more efficiently—reducing energy consumption, token processing, operational costs, and environmental impact. The following section connects prompt efficiency to broader ecological and social considerations.

1.3 Environmental and Social Impacts

The environmental cost of large-scale artificial intelligence is no longer an abstract concern; it is a tangible reality. Each interaction with a language model consumes electricity, requires data center cooling, and contributes to carbon emissions. Vague, inefficient, or excessively lengthy prompts amplify these impacts. Every excess token processed directly translates into higher energy consumption, increased server time, and a greater ecological burden.

This relationship between prompt design and environmental impact is grounded in the model's architecture. A longer prompt typically results in a longer output, which consumes more tokens. Data centers, often powered by non-renewable energy sources, consume computational resources with every token they process. While the environmental cost of a single prompt may appear negligible, the cumulative effect, across billions of daily interactions worldwide, makes prompt design a matter of real planetary significance.

Improving prompt structure is one of the most immediate and accessible ways to mitigate this cost. When users provide concise, well-scaffolded prompts, models require less processing power to arrive at a helpful answer. Optimizing prompt design streamlines processing demands, progressively decreasing energy usage and associated carbon emissions. Optimized prompting is not only a matter of performance; it is a low-barrier intervention with measurable sustainability benefits.

Social ones match these environmental stakes. As demand for AI continues to rise, access to model capacity becomes a contested resource. Public users, educators, researchers, and small organizations often face throttled access, longer wait times, or usage limits, especially during peak demand. Inefficient prompts monopolize system resources, indirectly restricting others' access and fostering digital scarcity, where a minority's practices limit the majority's opportunities.

Responsible prompt engineering, therefore, contributes to a more equitable and sustainable distribution of AI resources. Just as developers have long practiced energy-aware coding in software engineering, prompt engineers must adopt energy-aware prompting as a professional norm. Designing clear, efficient prompts is not only an individual skill; it is a social responsibility. By reducing excess processing and promoting shared access, prompt engineers help ensure that AI systems serve the broader public interest, rather than amplifying digital inequity or ecological harm.

1.4 The Ethical Obligation: Use It Right

As large language models become embedded in education, healthcare, legal systems, finance, and public administration, the stakes of AI use grow higher. The question is no longer whether people will use these tools, but whether they will do so responsibly. The ethical obligation to prompt well arises from the simple reality that prompts shape outcomes. A poorly crafted input can lead to inaccurate, misleading, or even harmful results, not because the model is malicious, but because it follows the user's instructions with statistical rigor and no external judgment.

Most professional fields couple powerful tools with structured guidance. Pilots, for instance, adhere to checklists. Engineers work within specifications. Medical practitioners adhere to protocols. These norms exist not merely to optimize performance, but to prevent harm. However, in prompting, a practice that governs how AI systems interpret and respond to human input, there remains no widely adopted standard. This absence is not a technical gap alone. It is an ethical one.

Unstructured prompting leaves room for misinterpretation. Vague queries risk generating hallucinated facts, while overly general requests may yield biased generalizations, which can be hazardous in high-stakes applications such as legal summarization, public health advisories, or personal data analysis. Errors may go unnoticed until they cause damage. More subtly, unstructured inputs reinforce a culture of irresponsibility, in which users are neither taught nor encouraged to reflect on how their instructions influence AI behavior.

The ethical responsibility of prompting extends beyond individual use. On the development side, ambiguous prompts undermine model alignment and safety testing. When researchers evaluate systems using inconsistent instructions, it becomes difficult to measure performance or diagnose failure modes. On the deployment side, untrained users may reproduce structural inequalities, amplify misinformation, or misuse systems in ways that escape accountability.

This book introduces a structured framework to address this challenge—not as a rigid moral code, but as a flexible design aid. Its purpose lies in helping users clarify intent, establish context, and reduce ambiguity. When practitioners systematically incorporate task parameters, audience considerations, tone, scope, and constraints, they achieve more precise outcomes, both technically and ethically. The approach fosters transparent, accountable interactions where human intentions and machine responses align more reliably.

Prompts are not neutral inputs. They are communicative acts with consequences. Treating prompt engineering as a professional practice, with its standards, norms, and expectations, elevates it from casual experimentation to civic responsibility. In this sense, using AI well is not only about fluency. It is about integrity. A brief ethics checklist later in the book offers practical guidance for reviewing prompts with attention to safety, accuracy, inclusion, and responsible use.

1.5 The Communication Gap: Human vs. AI

Human communication is rich, layered, and heavily contextual. We convey meaning not only through words but through tone of voice, facial expressions, body language, timing, and shared cultural references. Even a short phrase like *"Are you sure?"* can express doubt, encouragement, sarcasm, or concern, depending on how it is delivered and received.

Large language models do not operate in this world. They do not see, hear, or intuit context. They process only text, one token at a time, without access to human intent unless it is explicitly encoded. A prompt that leaves the audience unstated, the format ambiguous, or the tone undefined will not trigger clarification or empathy. It will trigger inference, based on probabilities drawn from the model's training data. The output achieves linguistic fluency but lacks semantic precision, not due to design failure, but rather because of inherent constraints, with tangible consequences for human-AI collaboration. Where a human reader might infer urgency from sentence structure or tone from punctuation, an LLM treats each token as part of a statistical sequence. It does not fill in context from shared experience or intuition. It responds only to what is present in the text.

This mismatch introduces risk. Users often assume that the model "understands" their request in a human sense. They phrase questions informally, omit key details, or rely on implication. The model, in turn, generates plausible responses that may be incomplete, misaligned, or subtly distorted in domains where clarity matters, such as healthcare, policy, or education. These subtle mismatches can lead to poor decisions or misinformation.

Prompt engineering addresses this gap not by developing more innovative models, but by providing more transparent communication. By structuring prompts intentionally, specifying role, audience, task, tone, format, and constraints, users reduce ambiguity and improve interpretability. They shift from assumption to articulation. Mastering this approach forms the cornerstone of prompt literacy, the skill of formulating instructions that large language models can consistently interpret and execute effectively.

Prompt literacy is not a niche skill; it is a vital one. It is becoming a prerequisite for working effectively with AI systems. Just as digital literacy became essential during the rise of the internet, prompt literacy is now essential in the age of generative models. It enables users not only to get better results but to interact with AI systems more safely, transparently, and responsibly.

Prompting, then, is not just input. It is translation, between the complex richness of human communication and the mechanical clarity required by machines. The more effectively we can make our intentions explicit, the better we can guide AI systems toward outcomes that serve human needs.

1.6 Previous Literature and Contributions

Prompt engineering has rapidly evolved from a fringe skill to a recognized field of inquiry, with contributions emerging from research labs, industry practice, and independent experimentation. Although the discipline is still young, a growing body of literature has begun to map its contours, offering taxonomies, design frameworks, and guiding heuristics.

One of the earliest systematic efforts came from industry practitioners seeking to formalize prompt design for real-world deployment. Among these, early frameworks based on job interview techniques, such as the STAR method, or analytic tools like SWOT, demonstrated how domain-specific structures could improve prompt precision. These approaches brought clarity to otherwise vague instructions and highlighted the role of formatting, sequencing, and context in shaping model behavior.

Academic contributions soon followed. In 2023, a widely cited *Prompt Pattern Catalogue* introduced recurring prompt types, such as few-shot, Chain-of-Thought, and self-reflective prompting, as reusable structures. Drawing on software engineering traditions, it treated prompts as design patterns that could be adapted, reused, and evaluated. Around the same time, the *Prompt Canvas* emerged as a visual tool for decomposing and assembling prompts, enabling users with varying technical backgrounds to understand the internal anatomy of a well-structured query.

By 2024, comprehensive surveys began to consolidate the field. These overviews documented advances in controllability, alignment, prompt-based fine-tuning, and interpretability. Importantly, they highlighted the absence of a shared framework that could span multiple domains, including software development, creative writing, education, healthcare, and legal analysis. Most existing tools, while useful, were designed either for technical experts or for narrow application areas, leaving a wide gap in accessibility and generalizability.

This book builds on that foundation by offering a unified and accessible structure: the Prompt Anatomy Blueprint. While earlier works provided valuable building blocks, they often assumed domain-specific knowledge or relied on implicit expertise. This framework serves explicitly interdisciplinary learners, policy professionals, educators, and self-taught practitioners by providing a consistent, transferable methodology for prompt design.

The book develops an adaptable vocabulary and structure for cross-domain application rather than a rigid universal solution, framing prompting as a human-centered practice that negotiates clarity, control, ethical foresight, accessibility, and adaptability. The Prompt Anatomy Blueprint formalizes this balance and provides a conceptual bridge between scattered techniques and a coherent professional standard.

1.7 A Conceptual First Look: Prompt Anatomy

Before formalizing any framework, it helps to examine what makes a prompt effective in practice. Consider the following instruction to an AI assistant:

> *The following executive summary, prepared by a senior sustainability consultant for the board of a UK-based energy company, adopts a formal tone while outlining three principal climate risks projected to affect operations by 2030. Use bullet points and limit each point to two sentences. Do not include a preamble or closing remarks. Write directly and clearly.*

Although it may appear simple, this prompt encodes a rich set of design decisions. It specifies the *role* (senior sustainability consultant), the *audience* (a corporate board), and the *task* (writing an executive summary). It imposes a *tone* (formal), a *structure* (bullet points), a *content constraint* (three risks), and a *length limit* (two sentences per point). It also excludes undesired elements, like introductions or conclusions. Each instruction nudges the model toward a more aligned, focused, and reusable response.

This internal structure is what we refer to as *prompt anatomy*. Like the anatomy of a well-formed sentence or a professional email, a prompt has identifiable components that shape its performance. While users often write prompts intuitively, high-performing prompts are rarely accidental. They are designed, sometimes implicitly, sometimes consciously, through choices about scope, format, tone, and audience.

Prompt anatomy does not require users to follow a rigid template. Instead, it offers a way to analyze and build prompts with purpose. It provides a shared vocabulary to describe what a prompt is doing and why. Just as one examines a paragraph's topic sentence, supporting details, and transitions, one can evaluate a prompt through its defined roles, imposed constraints, and structural logic.

This concept also clarifies why some prompts fail. When users omit key contextual elements, such as the intended audience, required format, or tone, the model compensates by drawing on its training data. The result may be plausible but off-target, verbose, or even misleading. Understanding prompt anatomy allows users to avoid these pitfalls and take greater ownership of the model's outputs.

In the subsequent chapters, we develop this concept further by presenting a practical framework that systematizes these elements into a workable structure. Each component corresponds to a key dimension of prompt design that practitioners can adjust, integrate, or limit according to their specific requirements.

Prompting, in this light, becomes a form of intentional design. Rather than relying on luck or trial and error, users can approach prompts as crafted tools, shaped to match a specific purpose, audience, and context. This shift, from improvisation to design, is the foundation of prompt literacy. A visual schematic of the Prompt Anatomy Blueprint later in the book illustrates each element and supports rapid practical adoption across domains.

1.8 Why Structure Matters

A prompt is not simply a message; it is a control mechanism. The structure of a prompt shapes the behavior of the model, influencing not just what the model says, but how accurately it aligns with the user's intent. When users approach prompting casually, they leave key elements unstated. The model must deduce the task, audience, tone, and output format, often producing plausible yet inconsistent, verbose, or misaligned responses.

Structure counteracts this ambiguity. By embedding purpose, constraints, and formatting into the prompt itself, users shift the burden from the model to the design of the interaction. For example, a prompt that specifies a tone, "use a formal, concise tone suitable for policymakers," provides more than a stylistic cue. A well-structured prompt directly influences the language, scope, and depth of the model's response, while defining a target audience or specifying numbered steps sharpens its reasoning and eliminates superfluous content.

Beyond technical benefits, structured prompting enhances *reliability*, and clear formatting yields predictable outputs that users can more easily assess, verify, and improve. It supports *reproducibility*: others can reuse or adapt the same prompt for related tasks without needing to start from scratch. Moreover, it facilitates *accountability*: when a prompt includes explicit choices, such as excluding sensitive topics or defining word limits, users gain a traceable record of their design intent.

Just as professionals in law, medicine, or software rely on documented procedures and standards, prompt engineers benefit from shared structures. A well-structured prompt functions as a blueprint or protocol, embedding decision-making in an accessible, transparent, and teachable format that transforms private experimentation into collaborative practice.

Structure also serves a normative function. It prompts users to consider the *ethical* and *social* consequences of their design choices. Specifying the intended audience, tone, or domain can surface questions of bias, accessibility, or appropriateness that might otherwise go unnoticed. A prompt written for students will differ, in tone and content, from one intended for legal professionals. Embedding these distinctions structurally promotes more equitable, relevant, and trustworthy outputs.

In short, structure is not decoration; it is foundation. It supports not only technical performance but ethical clarity, pedagogical reuse, and cross-domain applicability. As prompting becomes a standard skill across disciplines, structured design offers the most reliable path to responsible and effective use. It transforms prompts from mere requests into operational artifacts: designed, deliberate, and aligned with human goals.

1.9 Prompting As an Art and Science

Prompt engineering occupies a unique space between logic and expression, precision and intuition. Like programming, it involves structured syntax, control over variables, and clear input/output logic. Like writing, it relies on tone, audience awareness, and rhetorical framing. A short sidebar later in the book offers a concise restatement of these ideas for general readers without altering the conceptual argument. Treating it solely as a technical task risks overlooking its communicative and ethical dimensions. Treating it purely as creative writing risks overlooking measurable effects on performance and efficiency. Prompting, done well, requires fluency in both modes.

The scientific side of prompting is grounded in *reproducibility* and efficiency. Structured prompts allow for benchmarking, iteration, and optimization. A well-designed prompt can be reused across tasks, yielding consistent outputs under similar conditions. Practitioners can measure improvements, reduce token usage, and systematically refine their designs to achieve optimal results. These properties make prompting suitable for integration into software development cycles, educational workflows, and professional documentation standards. In this context, prompting becomes a form of lightweight programming, a way of encoding intention in a format the machine can reliably execute.

The artistic dimension lies in how prompts express voice, establish context, and frame interaction. Tone is not a cosmetic choice; it shapes the entire trajectory of the model's response. In sensitive domains such as healthcare, education, or cross-cultural communication, tone can be the difference between helpful and harmful. A prompt that fails to account for audience, nuance, or emotional framing may produce technically correct output that is nevertheless misleading, inappropriate, or ineffective. Prompting, therefore, draws on human insight as much as it does on model capacity.

Skilled prompt engineers recognize that design choices are rarely neutral. Every word, phrase, or instruction encodes assumptions about purpose, audience, and norms. A professional framework for prompt design becomes essential, as every conscious or unconscious decision directly influences the model's output and behavior. It provides a structured lens through which to examine, refine, and justify those decisions, without collapsing prompting into a rigid formula. It preserves creative flexibility while enforcing logical clarity.

Practitioners rarely achieve effective prompts on their first attempt; instead, they refine them through iterative cycles of testing, revision, and critical reflection. As with any design process, iteration reveals not only what works but why it works. Repeated practice develops predictive judgment, the capacity to foresee model responses and craft prompts strategically. This approach defines prompt literacy: merging technical understanding with communication expertise to generate ethically sound, contextually relevant, and dependable outcomes.

Prompting requires a dual approach, combining both art and science, as it merges interpretive understanding with methodological precision. Just as writing education balances grammar with voice or engineering education pairs calculation with design thinking, prompt engineering must develop both rigor and imagination. Even as models become more capable, their responses still depend on explicit instructions, making structured prompting necessary for clarity, safety, and reliable control. This book aims to provide the tools for both: a formal structure and the critical awareness to adapt it.

1.10 Prompting Across Domains

Research labs and technical teams no longer exclusively practice prompt engineering. It is now a practical necessity across a wide range of professional, educational, and civic settings. From courtrooms to classrooms, from hospitals to government offices, people are turning to large language models to assist with complex tasks, automate routine processes, and support decision-making. As the scope of AI use expands, so does the need for a standard that can prompt adaptation across domains without becoming overly abstract or prescriptive.

In healthcare, for example, clinicians may use language models to draft patient summaries, generate research overviews, or explain medical conditions in plain language. A prompt must be accurate, concise, and sensitive to patient needs. An overgeneralized instruction, such as "Explain this condition to a layperson," can result in vague or condescending output. A structured prompt that specifies the intended reading level, tone, and purpose ensures greater clarity and trust.

In education, prompting is increasingly used for curriculum design, formative assessment, and interactive tutoring. Teachers and students rely on models to generate lesson plans, provide feedback, or simulate Socratic dialogue. Creating effective educational prompts requires striking a balance between instructional quality and strict adherence to academic benchmarks while simultaneously crafting content that resonates with students' developmental levels and captures their interest. When educators design prompts that incorporate specific learning objectives and account for learners' prior knowledge, the resulting outputs become significantly more relevant and impactful.

In law, prompting plays a role in summarizing case law, drafting arguments, or interpreting legal language. The stakes are high: ambiguity or hallucination can lead to serious misunderstandings or procedural errors. Legal prompts require precision, neutrality, and awareness of jurisdictional context. When a lawyer prompts a model to "summarize this case," the result varies dramatically depending on whether the jurisdiction, intended audience, and formatting are made explicit.

Even in creative industries, where flexibility is highly valued, structured prompting enhances consistency in output and creative control. Writers, journalists, designers, and filmmakers utilize language models to generate ideas, reframe narratives, or draft content in a specific voice. Without clear guidance, the model may offer clichéd or genre-inappropriate responses. With prompt components like tone, format, or narrative constraints, creative professionals can treat the model as a reliable collaborator, not a wildcard.

What unites these domains is not the type of output, but the nature of the interaction: each relies on human judgment to specify intent and define expectations. Without a shared language for doing so, users must improvise, repeating mistakes or relying on undocumented intuition. A professional prompting framework can offer this bridge, providing a flexible structure for clarifying purpose, tone, audience, and output format without being overly rigid. Such a structure must apply across domains while allowing for local adaptation. It does not prescribe content; it clarifies design.

As AI systems become an integral part of daily work across various professions, the ability to write effective prompts will become a foundational skill. Whether in hospitals, classrooms, courts, or studios, prompting must be teachable, transferable, and ethically grounded. This book builds toward a unified framework to meet that need, not to constrain creativity or expertise, but to provide a shared foundation for responsible prompting in real-world contexts.

The more widely language models are adopted, the more critical it becomes to build prompting skills into everyday professional literacy. Whether in a hospital, a university, a law firm, or a creative studio, prompting is becoming a core competency. A unified framework ensures that this competency is both teachable and transferable and grounded in ethics.

The book constructs a core framework that maintains creative flexibility and field-specific expertise while instituting universal principles for ethical prompt design, addressing AI's expanding role beyond specialized domains. Even with rapidly advancing models, effective results still depend on explicit design choices, reinforcing the need for prompt literacy rather than reducing it. A short comparative prompt table later in the book contrasts domain-specific priorities across healthcare, education, law, and creative practice, illustrating how roles, tone, and common constraints shift while the underlying structure remains constant. Prompting across domains is not a future challenge. It is today's reality.

1.11 Summary

This chapter established the case for prompt engineering as a professional standard rather than an informal practice. It showed that vague or ad hoc prompting produces unnecessary computational cost, environmental burden, and increased risk of hallucination, bias, or misalignment in high-stakes domains. By analyzing practical and ethical consequences across education, healthcare, law, policy, and creative industries, the chapter argued that prompting now constitutes an essential digital literacy.

The chapter traced why structured prompting matters: it reduces token waste, improves reliability, and strengthens alignment with human intentions while promoting equitable access to AI capacity. It framed prompting as a communicative design discipline grounded in responsibility, clarity, and sustainable use of advanced language models. Through examples and literature context, the chapter positioned prompt engineering as both an emerging art and a measurable science that now requires shared norms and standards.

CHAPTER 2

Prompt Engineering for Everyone

2.1 Why Prompt Engineering Matters Now

The discipline of prompt engineering has emerged in response to a fundamental shift in human–computer interaction. Earlier computing paradigms required users to learn programming languages, follow structured commands, or work through graphical interfaces to engage meaningfully with machines. In contrast, modern systems powered by large language models (LLMs) now respond directly to natural language instructions. This transformation opens up new possibilities for interaction, but it also demands a new kind of intentionality. The burden of precision has moved from code to language, from programming to articulation.

This change is not merely technical. It repositions the user as an active designer of AI behavior. Instead of writing scripts in Python or configuring command-line parameters, a professional might now compose a prompt such as: *"You are an academic coach. Summarize this student essay in three bullet points using encouraging, supportive language."* In this prompt, the structure, tone, role, and task are embedded entirely in natural language. The language itself becomes the interface.

H. Tavakoli, *Prompt Engineering for Everyone*,
https://doi.org/10.1007/979-8-8688-2338-1_2

Prompt engineering transforms the relationship between people and intelligent systems. In previous decades, directing a computer to take action required adherence to formal syntax and often relied on technical intermediaries. Today, professionals in fields such as education, journalism, medicine, and policy can articulate their goals in everyday terms and receive structured, context-aware responses. The complexity of design has not disappeared; it has merely shifted. The interpretive responsibility now lies with those crafting instructions, not with those writing code.

This shift carries broader implications. Prompt engineering signals the emergence of a new form of digital literacy, one that shapes how individuals pose questions, assess responses, and assign accountability in automated systems. Much like email redefined professional correspondence and search engines transformed access to information, prompting a reconfiguration of how people engage with knowledge and decision-making.

In this light, prompt engineering is not a passing technical trend. It is an emerging literacy that shapes how people communicate with and through intelligent systems. As the next section will show, its foundations are both practical and conceptual, and they begin with understanding what prompting truly is. A brief reminder in later chapters links each CASTROFF element to responsible use, emphasizing accuracy, inclusivity, and caution when prompts influence high-impact decisions.

2.2 What Is Prompt Engineering?

Prompt engineering is the deliberate composition of natural language instructions designed to guide the behavior of a large language model. It is not simply the act of asking a machine to act. It is a structured form of authorship, where language functions as a control surface through which users shape system output. A prompt specifies not only the task but also tone, scope, format, constraints, and suppressions. Its goal is not just coherence but alignment: between intention and output, between user goals and system behavior.

The performing arts offer an instructive analogy: if we consider the language model a highly skilled but literal actor, the prompt serves as the script. A vague script frequently yields a flat performance. A precise one elicits clarity, tone, and nuance. Like a playwright who shapes the actor's timing, stance, and delivery through language alone, the prompt engineer shapes the model's voice, reasoning, and emphasis.

Viewed this way, prompt engineering is more than a technical trick. It is a design practice. The system embeds assumptions, foregrounds values, and influences downstream user experiences. In settings such as education, clinical guidance, journalism, or public policy, the prompt becomes inseparable from the legitimacy of the outcome. When a language model rephrases a medical report for a patient or summarizes legal advice for non-experts, the prompt's phrasing influences not only understanding but also trust, action, and accountability.

Prompt engineering does not belong solely to computer science. Its relevance extends to anyone who guides, reviews, or relies on AI-generated output. It requires precision without code, structure without formal syntax, and ethical awareness embedded directly in language. As such, it sits at the intersection of writing, design, reasoning, and responsibility.

The following section examines why this practice is not limited to technical experts, but rather extends to professionals and practitioners across every domain that involves human communication. In later chapters, short examples from education, healthcare, and policy illustrate how CASTROFF adapts across domains without altering its underlying structure. For example, medical summarization may require plain language for patients, while legal summarization emphasizes precision and citation awareness. These variations show how CASTROFF adjusts to domain expectations while retaining a consistent framework. Each element can also be applied with explicit attention to responsible use, including accuracy, inclusion, and clarity, developed further throughout the book. Later chapters briefly reference widely used prompt guidance that has informed the CASTROFF approach, without requiring readers to consult specialist sources.

2.3 Why Non-coders Should Care

Prompt engineering is no longer the exclusive concern of computer scientists or machine learning specialists. Across professions, people now interact with AI systems whose behavior is quietly, but profoundly, shaped by the language instructions behind the scenes. Whether reviewing model-generated case notes, designing AI-supported feedback systems, or working with auto-summarized regulatory guidance, non-technical professionals increasingly rely on outputs created by prompts they did not author and often cannot see.

This reliance spans fields as diverse as healthcare, education, journalism, law, policy, and design. A clinician may receive AI-generated patient histories. A teacher may assess student work shaped by automated feedback. A lawyer may review contracts summarized by natural language tools. In each case, the outcome appears polished, but the underlying behavior is governed by a prompt written elsewhere. Understanding how these prompts function is no longer optional. It is essential to interpret, validate, and utilize AI outputs safely in professional settings.

Prompting now acts as a structuring mechanism far beyond software development. The prompt shapes task framing, role simulation, and tone control. The risks are real. A summarization prompt that favors brevity over nuance may erase dissent. A feedback tool that standardizes tone may suppress identity. A translation system guided by culturally narrow norms may distort meaning for multilingual users. These are not isolated anomalies. Widely used tools that incorporate prompts amplify these effects, potentially creating systemic distortions even when the model functions as designed.

We must therefore recognize prompt engineering as an interdisciplinary literacy. It shares qualities with instructional design, interface writing, legal drafting, and editorial planning. It demands conceptual clarity, ethical attentiveness, and contextual awareness. Effective prompts do not just generate outputs; they set the terms of interpretation and action.

Prompt engineering belongs to everyone who relies on AI to support decision-making, documentation, communication, or care. It is not a technical specialism but a shared responsibility. By treating prompting as a form of civic and professional literacy, this book invites readers from all disciplines to approach it with intention, reflection, and purpose. Later chapters include short notes on responsible use, showing how CASTROFF elements such as tone, audience, and constraint settings help reduce unintended ethical or cultural harm.

2.4 Why Now? Three Forces Behind the AI Boom

The emergence of prompt engineering as a foundational skill is not accidental. It reflects the convergence of three technological forces: data scale, computational infrastructure, and architectural innovation. Together, these developments have redefined the capabilities of language models and the expectations placed on their users.

The first force is the exponential growth of digital content. Modern language models differ from earlier language processing systems in that they learn from vast, unstructured collections, including encyclopedias, news archives, scientific literature, discussion forums, and code repositories, rather than from curated corpora. This breadth enables the models to generalize across domains and respond with remarkable fluency. However, it also increases the risk of redundancy, stereotype propagation, and epistemic flattening. In this environment, prompts act as steering mechanisms, defining the model's interpretive frame and constraining its expansive knowledge base.

The second force is the rise of advanced hardware. The availability of high-efficiency processors, such as GPUs, TPUs, and dedicated AI accelerators, has transformed the economics of model development. Training tasks that once took months on academic clusters are now complete in days. This acceleration has enabled rapid iteration, but it has also increased energy consumption and environmental impact. Within this context, prompt engineering becomes a form of computational stewardship. A well-designed prompt reduces processing time, minimizes token waste, and supports more sustainable AI usage at scale.

The third and most transformative force is architectural innovation, particularly the development of transformer-based models. Introduced in 2017, transformers enable models to process entire sequences of language simultaneously while attending to contextual relationships between words. This capacity for self-attention, rather than fixed proximity, allows for fluent, context-sensitive outputs across a wide range of tasks. It also makes prompting a viable primary interface. Rather than retraining a model for every use case, users can now articulate intent through natural language and guide the model's behavior in real time.

Together, these three forces, *scale, speed,* and *structure,* have made prompt engineering both possible and necessary. As AI systems become more accessible, the responsibility for directing their outputs no longer belongs solely to developers. It increasingly rests with the individuals and institutions that write the prompts, which shape what these models say, how they say it, and to whom they are accountable. This chapter briefly notes that efficient prompting can also reduce unnecessary token usage, supporting more sustainable AI interaction when working with large models.

2.5 How We Got Here: From Rule-Based AI to Transformers

To understand the value of prompt engineering in today's systems, it is helpful to place it within the historical arc of natural language processing. Over the past five decades, the field has progressed through three primary stages: rule-based logic, statistical modelling, and deep learning. Each stage has redefined how machines interpret language and how users must communicate with them.

The earliest systems relied on hand-coded rules. These rule-based approaches mapped specific phrases or grammar patterns to predetermined responses, often using decision trees or condition–action templates. While precise in narrow domains, they were brittle. Minor phrasing changes, such as the omission of an article or the reordering of words, could lead to complete failure. These systems could not adapt to ambiguity, dialect variation, or semantic nuance. Their rigidity made them unsuitable for open-ended dialogue or large-scale deployment.

The second era introduced statistical language models. Systems based on n-grams, TF–IDF, and hidden Markov models began to estimate the likelihood of word sequences and rank relevance using probabilistic methods. These models enabled greater flexibility than rules alone, capturing patterns from data rather than relying solely on predefined logic. However, they lacked depth. They treated language as surface-level text, often ignoring the broader context. The outputs were more fluent than those of rule-based systems, but often mechanical or awkward, grammatically correct yet lacking meaning.

Machine learning then emerged as a corrective. Earlier systems employed classifiers trained on labelled datasets to perform tasks like topic identification, entity recognition, and sentiment detection. These systems could generalize more effectively than earlier methods, but they still required handcrafted features and remained limited in scope. These systems could neither reason across extended passages nor interpret intention from naturalistic input.

The shift to deep learning, and especially to transformer-based models, marked a fundamental break from these traditions. Rather than processing words one at a time, transformers evaluate all input tokens in parallel. They utilize self-attention to model the relevance of each word to its context in the sequence, enabling them to resolve ambiguity, track long-range dependencies, and generate fluid responses tailored to the context. Crucially, they do not require fixed rules or labelled features. Instead, they learn relational structure directly from the data.

Prompt engineering becomes feasible only in this context. Transformer-based models are sensitive to the framing of input, tone, sequence, constraints, and clarity, all of which influence how the model interprets and completes a task. What would have once required reprogramming can now be accomplished through a well-crafted prompt. This shift in power, from system logic to user language, has created both new capabilities and new responsibilities.

Prompting is therefore not a workaround or a trick. It is the front line of communication with modern AI systems. Understanding its roots helps to clarify its function. Prompt engineering is the culmination of decades of progress in computational linguistics, human–machine interaction, and AI system design. It is where historical limitations give way to conversational possibility and where user intent becomes the primary instrument of control.

2.6 What Changed Everything: Transformers

The introduction of the transformer architecture marked the most decisive development in language model evolution. First described by Vaswani et al. in 2017, the transformer replaced the step-by-step limitations of earlier models with a structure that processes input tokens simultaneously, using attention mechanisms to determine relevance. This architectural shift transformed the scale, speed, and quality of natural language understanding.

At the center of this innovation is self-attention. Transformers differ from recurrent or convolutional models by analyzing relationships between all words in a sentence simultaneously rather than processing language linearly. This approach enables the model to assess grammatical and contextual cues in ambiguous sentences, such as determining whether "the telescope" refers to "the lion" or "the man" in the phrase "The lion saw the man with the telescope," by weighing competing interpretations. These systems do not merely count word frequency; they reason across relational cues.

Positional encoding is another core feature. Transformers rely on added positional encodings to maintain sequence information since they lack inherent word-order processing. The mechanism enables the model to recognize the difference between structurally contrasting phrases such as "The lawyer questioned the witness" and "The witness questioned the lawyer," despite using the exact words. In this way, the transformer architecture retains the ability to interpret both structure and meaning across long passages.

The benefits are not limited to comprehension. Transformers also introduced architectural parallelism; earlier models processed language sequentially, which slowed both training and inference. Transformers process input in parallel instead, enabling them to scale to billions of parameters and train on massive datasets. This scalability has enabled models to generalize across domains, transfer skills between tasks, and respond with high fluency and contextual sensitivity.

These changes make prompt engineering possible. Because transformers interpret input holistically, they can adapt to nuanced prompts that specify task, tone, format, or constraints. The model does not require new code to switch from legal summarization to poetic rewriting;

it requires a well-phrased instruction. The prompt, in this sense, becomes the user's programming language. It controls what the model attends to, how it frames the task, and what constraints guide the output.

Prompting, therefore, is not a workaround or an improvisation. It is the formal interface between human intent and machine capacity. The transformer enables this interface by making model behavior responsive to natural language structure. The prompt provides the structure; the transformer interprets it. This relationship is what makes prompt engineering both powerful and essential.

2.7 How Prompting Fits In

With the emergence of transformer-based models, prompting has become the principal interface between human intention and machine behavior. Just as search engines revolutionized information retrieval, large language models now transform knowledge generation, decision-making, and problem-solving, directed not by programmed code but by natural language prompts.

Prompt engineering, in this context, is no longer a peripheral skill. It has become a core digital literacy. Whether a user is drafting a policy memo, reviewing medical guidance, analyzing financial data, or providing legal aid, the ability to formulate clear, context-sensitive instructions now shapes the quality of the model's output. This shift extends beyond computational fluency: it requires users to think intentionally about how language encodes roles, assumptions, constraints, and desired outcomes.

Prompt construction now significantly influences the reliability and ethical foundation of AI responses. A vague request invites misinterpretation. An imprecise role assignment may distort tone. A missing constraint can lead to irrelevant or unsafe output. Each of these risks is avoidable, not by reprogramming the model, but by designing better prompts. Not every CASTROFF element is required in every prompt; low-risk or routine tasks may rely on only a subset, while high-stakes contexts benefit from more complete application.

We must therefore treat prompt engineering as both an expressive and technical discipline. It demands attentiveness to audience, clarity of task, control of tone, and sensitivity to social and institutional context. When done well, prompting makes machine responses more useful, human interactions more coherent, and system behavior more predictable. A brief ethical note accompanies each CASTROFF element later in the book, highlighting accuracy, inclusion, and domain-appropriate responsibility as essential components of structured prompting.

This chapter outlined the conceptual and historical foundations of prompt engineering, including its rise alongside transformer models, its relevance across various domains, and its emergence as a shared professional literacy. In addition, a simple schematic later in the book visually annotates an example prompt with all CASTROFF elements to support rapid comprehension. The next chapter introduces the *prompt anatomy* framework, a structured approach to composing prompts that are efficient, inclusive, and adaptable. By learning to recognize the components of a well-formed prompt, users can gain deeper control over generative systems and begin to shape their outputs with purpose and care.

2.8 Summary

This chapter explained why prompt engineering has become a central skill in modern AI use. It traced the shift from rule-based systems to transformer models and showed how natural language now functions as the primary interface for directing machine behavior. It argued that prompting is not a technical trick but an emerging form of digital literacy that shapes how users instruct, interpret, and take responsibility for AI-generated outputs. Across different fields and professions, prompt engineering is now a shared practice grounded in clarity, intention, and careful design.

2.9 Self-Assessment Quiz

1. Which of the following best describes the primary shift in human–computer interaction enabled by prompt engineering?

   a) Users now need to learn multiple programming languages to communicate with AI systems.

   b) Machines interpret vague instructions and infer user goals independently.

   c) Natural language prompts replace traditional coding, requiring clarity and strategic intent from users.

   d) AI systems automatically correct ambiguous user input without guidance.

2. Why has prompt engineering become a professional literacy beyond the technical domain?

   a) Because it is required only for building AI models from scratch

   b) Because professionals across many fields must now guide AI systems through natural language

   c) Because it eliminates all risks associated with AI bias

   d) Because it replaces all human decision-making

3. What are the three main technological forces that have made prompt engineering possible and necessary?

   a) Data scale, faster computation, and transformer architectures

   b) Programming languages, cloud storage, and mobile devices

   c) Rule-based logic, symbolic AI, and expert systems

   d) Graph databases, virtual reality, and blockchain

4. How do transformer architectures contribute to the effectiveness of prompt engineering?

   a) They process inputs sequentially, limiting context understanding.

   b) They use self-attention and parallel processing to capture relationships across the entire input.

   c) They rely exclusively on fixed word order without contextual adaptation.

   d) They generate responses based on random word selection.

5. Why is prompt engineering described as a new form of digital literacy?

   a) Because it requires memorizing complex coding syntax.

   b) Because it enables users to structure instructions clearly to guide AI output responsibly.

   c) This approach removes the requirement for human oversight in AI systems.

   d) Because it simplifies all AI tasks to a single universal command.

CHAPTER 3

Prompt Literacy: From Habits to Mastery

For most users, interacting with an AI model feels intuitive: type a sentence and receive a response. After all, this mirrors how we have used technology for decades. Whether searching the web, texting a friend, or talking to a voice assistant, the default mental model has been shaped by tools that prioritize brevity, keywords, and casual phrasing. Users enter a fragment, hit send, and expect relevance. These habits, reinforced by years of web search and autocomplete systems, feel natural, but they are no longer enough.

Large language models (LLMs) are not search engines. They do not scan databases for prewritten answers or fetch authoritative links. Instead, they generate outputs probabilistically, based on learned patterns from vast datasets. However, many users continue to engage with LLMs as if they were querying Google or chatting with a help desk bot. They type: "Tell me about climate change?" or "Can you write a paragraph on AI?" Users frequently provide unstructured, context-poor inputs not through carelessness, but because previous systems have conditioned them to prioritize speed over precision.

H. Tavakoli, *Prompt Engineering for Everyone*,
https://doi.org/10.1007/979-8-8688-2338-1_3

This behavioral mismatch leads to a frustrating paradox: intelligent models producing mediocre outputs. When the prompt is vague, the model makes a guess. When the instruction is minimal, the model elaborates unnecessarily or veers off topic. Users may interpret this as a flaw in the AI itself, rather than a reflection of inherited prompting habits that no longer fit the medium. The real issue is not the model's capacity; it is our approach to it.

Prompt engineering begins by acknowledging this shift. It asks users to recognize that LLMs do not read minds; they simulate meaning. Therefore, communication must become intentional, structured, and tailored to the machine's interpretive logic. The habits that served us well in keyword searches or casual messaging, such as minimalism, ambiguity, and passivity, must give way to clarity, constraint, and active instruction.

For example, a user asking, "How does inflation work?" may receive a generic explanation. However, if they specify, "Explain how inflation affects mortgage holders in the UK in under 100 words," the result is sharper, more relevant, and easier to assess. This transformation is not a technical upgrade; it is a literacy upgrade. Prompt literacy.

This chapter is about that transition. It begins with the habits we have inherited, then builds a practical understanding of how to write for LLMs with intent, structure, and adaptability. Prompting is not about coding or tricking the machine. It is about learning a new way to write: one that makes human goals legible to systems that do not understand us, but can nonetheless act on our instructions, if we learn to speak their language.

3.1 New Systems, New Rules

Effective prompting requires a fundamental understanding of the system in question. Many users mistakenly treat large language models (LLMs) as equivalent to chatbots, voice assistants, or search engines, when in fact they operate on fundamentally different principles. These systems, while superficially similar, are built on fundamentally different architectures, logics, and user assumptions. Treating LLMs like chatbots risks misunderstanding both their strengths and their limitations.

Rule-based chatbots, which have dominated digital interfaces for decades, are essentially scripted decision trees. They recognize specific keywords or commands and return predefined responses. When a user types "reset my password," a rule-based system triggers a fixed pathway: verify identity, present options, and deliver the next step. These systems are efficient for narrow tasks, but brittle when confronted with ambiguity, nuance, or complexity. Their intelligence is artificial in the most literal sense; it is entirely manufactured by the system designer.

LLMs, by contrast, are probabilistic sequence predictors trained on massive corpora of text. They do not follow scripts. They do not retrieve stored responses. Instead, they generate output dynamically, one token at a time, based on statistical associations in their training data. When prompted, they calculate the word that is most likely to come next, then repeat this process until the response is complete. This process can simulate humanlike reasoning, but it does not involve comprehension, memory, or intentionality.

The shift from rules to probabilities introduces new challenges and new opportunities for users. LLMs transcend the constraints of pre-programmed instructions that limit rule-based systems. They can compose poetry, translate between languages, simulate expert debate, or generate legal drafts. However, this flexibility comes at a cost: unpredictability. Because there is no fixed knowledge base, the quality of output depends almost entirely on the structure, clarity, and specificity of the input. The system cannot request clarification or identify ambiguity unless specifically instructed to do so. This fundamental limitation explains why effective prompting constitutes a programming discipline rather than casual conversation. Each prompt is a command sequence written in natural language, shaping the model's generative trajectory.

The user is not conversing with a sentient assistant but designing a language-based interface for controlled output. An underspecified prompt, such as "Write something about climate change," causes the model to revert to safe, generic response patterns. When the prompt is over-determined, e.g., "Write a 150-word briefing for a UK policymaker outlining the social effects of rising sea levels, using plain language and one real-world example," the model performs with greater relevance and coherence.

Understanding this new system logic changes the user's responsibility. In legacy systems, failure often rested with the software: it could not handle variability, and so users learned to simplify their inputs. In LLM systems, failures often occur when the prompt is vague, unscaffolded, or mismatched to the model's capabilities. The more precise the prompt, the more useful the output will be.

In short, prompting an LLM is not casual interaction; it is structured instruction. Users must unlearn the habits of chat and search and instead adopt a mindset closer to that of a systems designer. The model is not a human. It is not a help desk. It is a mirror of language, shaped by pattern, structure, and probability. To use it well, we must learn to speak its logic.

3.2 The Shift to Prompt Literacy

In the transition from casual AI use to intentional prompt engineering, a more profound shift is taking place, one that mirrors other literacy revolutions in human history. Just as reading and writing required new ways of thinking when the printing press emerged, large language models demand their form of structured literacy. This literacy is not about technical coding or algorithmic knowledge; it is about writing clearly and strategically for a machine that simulates language, but does not understand it.

Prompt literacy refers to the ability to communicate effectively with an artificial language model by shaping the form, tone, structure, and constraints of one's request. It involves more than just phrasing a question well. It means understanding what the model *needs* in order to behave as the user intends. Unlike a conversation with a person, who can ask clarifying questions, guess your intention, or adapt to ambiguity, a large language model behaves like a probability engine. It completes the user's prompt with what is most statistically likely to come next, unless the user intervenes with well-crafted instructions.

Developing prompt literacy requires a shift in mindset. Many users approach language models the way they approach Google Search or voice assistants. They type a short, vague query and expect the system to infer their intended meaning. Others treat it like a chatbot, relying on an informal or conversational tone without structure. These habits may yield usable results for simple tasks, but they break down quickly in complex, sensitive, or professional scenarios. The burden of clarity in prompting falls entirely on the user.

This shift is both cognitive and linguistic in nature. It asks users to think like instructional designers, not conversationalists. Prompts must define the task, the audience, the expected format, and any exclusions. They must anticipate failure modes, such as hallucinations or generic answers, and pre-empt them with specificity. In doing so, prompt literacy begins to resemble a craft: part language use, part interface design, and part strategic thinking. It is a literacy of *simulation control*, where the user learns how to shape the model's behavior without access to the underlying code.

Significantly, this shift extends beyond engineers, developers, and technical experts to a broader audience. Anyone who interacts with AI systems, including educators, policymakers, researchers, and artists, can benefit from developing this literacy. As language models become embedded in tools, platforms, and workflows, the skill of writing effective prompts becomes as foundational as using email or editing a document. Prompting is not just a technique; it is a mode of authorship in the age of machine-generated language.

Mastering this new literacy requires exposure, reflection, and practice. It also requires a new vocabulary, a shared language for describing different prompt types, output styles, and system behaviors. In the following sections, we build this foundation, not by prescribing rules, but by introducing the concepts and strategies that make prompting intelligible, testable, and improvable. Prompt literacy begins with awareness, but it matures through design.

3.3 Linguistic Foundations: Clarity, Order, and Scope

Large language models simulate meaning through statistical prediction, not understanding. This simple fact has profound implications for how we should write prompts. Statistical language models lack the human capacity for inferring intent, resolving ambiguity, or supplying missing context unless their training data reinforces explicitly these patterns. These systems generate responses based on probabilistic patterns in their training corpora rather than comprehending meaning as humans do. Consequently, effective prompting requires meticulous attention to three linguistic dimensions: *clarity* of expression, logical sequencing, and *precise scope* definition.

Clarity refers to how explicitly a prompt communicates its intent. Human readers often tolerate vague instructions through shared context or emotional intuition; machines do not. For example, a request like "Summarize this" provides no detail about audience, tone, format, or length. The model may respond fluently, but its output will vary widely depending on implicit biases, prior examples in its training data, or even randomness in the generation process. In contrast, a clear prompt, such as "Write a 100-word summary in plain English for a high school audience, focusing only on the main argument," provides constraints that reduce this variability and make the output more useful.

The sequence of information presentation, whether through word order, task order, or even punctuation, directly affects the model's responses, particularly when processing multi-step instructions. For instance, "Write a conclusion and then explain how you got there" produces a different result than "First explain your reasoning, then write a conclusion." Models are sensitive to structural cues because they process text as a chain of dependencies rather than a hierarchy of ideas. Disordered prompts often yield incoherent or truncated outputs, especially when tasks are complex or involve nested structures.

Scope refers to the boundaries of the task. Many prompt failures occur when users request too much at once or fail to specify what should be included or excluded. Consider a prompt like "Describe the economic, political, and cultural consequences of globalization." This instruction may seem reasonable to a human, but to a model, it is vast and underspecified. Without constraints, such as word count, time frame, or region, the model may drift, bloat, or latch onto superficial generalizations. Effective prompts define their perimeter: what the model *should* cover and equally what it *should not*.

Human prompters face the challenge that natural language frequently contains ambiguity, nonlinearity, and rich context—characteristics that strengthen human dialogue but create difficulties for statistical models. Prompting thus requires a disciplined use of language, what linguists might call *low-ambiguity, high-instructionality discourse*. Prompts must avoid pronouns without referents, modifiers without anchors, or terms that depend on shared human knowledge rather than machine-learned associations.

Pedagogically, this section reinforces a crucial shift: prompting is not casual writing. It is not expressive, poetic, or speculative unless specifically instructed to be. It is a form of *instructional design*, and as such, it demands the same linguistic discipline we expect in legal drafting, technical writing, or procedural documentation. Clarity ensures the model interprets the task correctly. Order ensures the model executes it coherently. Scope ensures the model stays on track.

Together, these three dimensions form the core of prompt literacy. They do not guarantee perfect results, no prompt ever does, but they maximize the alignment between human intent and machine output. When neglected, even the most sophisticated AI system will revert to shallow pattern-matching and generic responses. When applied with care, these principles transform a prompt from a question into a framework, from a suggestion into an executable instruction.

3.4 Prompt Types: One-Shot, Role-Based, Instructional

Throughout this chapter, "pre-prompt" refers to text that sets context before the main instruction, while "follow-up prompt" is any subsequent instruction used to refine, expand, or correct an initial answer. One of the most effective ways to become proficient in literacy is to understand the structural types of prompts that shape a model's behavior. Not all prompts serve the same function, and not all of them guide the model with the same degree of precision. While countless variations exist, three basic types of prompting have become foundational: one-shot prompts, role-based prompts, and instructional prompts. Each offers a distinct affordance, and learning when and how to use them is an essential skill for anyone working with language models.

One-shot prompting refers to the most straightforward approach to this task. It involves presenting a single example, request, or template and asking the model to continue in the same manner. For example, if one writes, "Translate the following sentence into French: 'I am learning AI,'" and stops there, the model is likely to produce a translation immediately. This kind of prompt is fast, compact, and efficient, but it relies on implicit context. The model assumes action based solely on pattern recognition. As a result, one-shot prompting works well for straightforward tasks such as translation, summarization, or sentence correction. However, it often falls short in cases that require nuanced understanding, constraints, or audience adaptation.

Role-based prompting introduces an additional layer of specificity by assigning the model a particular identity or point of view. For instance, a user might begin with, "You are a legal expert advising a non-profit organization" or "You are an empathetic therapist responding to a young client." These prompts help frame the model's tone, vocabulary, and perspective by activating relevant patterns from its training data. Role-based prompting proves particularly effective for outputs that require professional, stylistic, or ethical constraints. The user guides the model's responses by specifying its intended persona rather than giving direct instructions.

Instructional prompting, by contrast, frames the request in explicitly directive language. Instead of relying on example-based continuation or identity framing, instructional prompts provide the model with precise instructions on what to do. These prompts often include verbs such as *summarize, list, compare, rephrase, predict,* or *critique,* and they may contain multiple clauses that define the task, format, and intended audience. For example, a user might write: "Summarize the following article in three bullet points suitable for a policymaker unfamiliar with the topic. Use clear, non-technical language." This type of prompt functions like a command, often yielding highly structured and interpretable output.

Each of these prompt types offers strengths and limitations depending on the task at hand. One-shot prompts are fast but fragile. Role-based prompts are adaptive but context-dependent. Instructional prompts are explicit but risk verbosity or constraint violations if not carefully worded. In practice, many effective prompts combine these types; for instance, assigning a role and providing an instruction or offering an example followed by a task description. Understanding the grammar of each prompt type allows users to mix and adapt them fluently.

As users become more fluent in these structures, they gain finer control over the model's output. More importantly, they begin to see prompting not as a mechanical query, but as a rhetorical act, a way of shaping the relationship between the user, the model, and the desired outcome. Just as a skilled writer chooses between narrative modes or sentence types, a prompt engineer selects the appropriate prompt type to achieve the desired communicative goal. This fluency marks the difference between casual experimentation and strategic design.

3.5 Format and Style Control: Tone, Length, and Output Shaping

In any communicative exchange, the *what* is only part of the message. The *how*, tone, structure, length, and formatting play an equally vital role in determining whether an output is usable, appropriate, or persuasive. This principle applies equally when prompting large language models. If a prompt leaves these stylistic or structural expectations implicit, the model will generate an answer based on general statistical norms, which may not align with the user's intended meaning. If the prompt specifies these expectations, the model is more likely to produce an output that aligns with the desired communicative context. Learning to control format and style is, therefore, central to the practice of prompt engineering.

Tone control enables users to shape the emotional register, formality, and personality of the model's voice. For example, a medical professional may require the model to generate responses in a calm, reassuring tone when writing guidance for patients. In contrast, the same information delivered to a policymaker may need to sound neutral, technical, and authoritative. Phrasing like "Write in a formal tone" or "Use a conversational voice appropriate for teenagers" can radically alter how the model frames its language, metaphors, and rhetorical strategies. While the core content may remain the same, the tone affects how it is received and understood.

Length control, though seemingly simple, often requires careful calibration. If no length is specified, the model may err on the side of over-explanation or meandering elaboration, especially in low-stakes contexts where verbosity is statistically usual. However, in professional settings, grant proposals, press releases, and legal memos, every word matters. Prompts that state, "In 150 words or fewer" or "Use no more than three bullet points," help ensure that the model's output fits predefined constraints. This constraint is particularly useful in systems where the output will be used downstream, such as summarization for publication or automated formatting pipelines. By defining upper and lower bounds for response length, prompt engineers reduce drift and make outputs easier to audit, reuse, or embed.

Output format control involves defining the structural shape of the model's response: whether it should appear as a paragraph, a list, a table, a markdown document, or a structured JSON output. This becomes essential when integrating the model into workflows or applications. For instance, if a prompt specifies "Respond in a three-column table comparing risks, benefits, and unknowns," the model will attempt to format the output accordingly. Similarly, prompts that request "a numbered list of steps" or "a five-paragraph essay" invoke recognizable textual schemas that shape the model's generation logic. Format scaffolding serves as a structural boundary, reducing uncertainty and enhancing readability.

Together, tone, length, and format comprise the stylistic frame that surrounds a model's content. When these elements are made explicit in a prompt, the model's output becomes not only more usable but also more trustworthy, especially in contexts that demand professional rigor. This is where prompt engineering departs from casual interaction and becomes a tool of communicative design.

Mastery of format and style control also supports accessibility and ethical responsibility. Well-crafted prompts deliberately guide the model to tailor technical information for particular readerships, whether by simplifying explanations for young learners or employing accessible formatting for neurodiverse audiences, demonstrating how audience awareness fundamentally determines communication effectiveness. This approach transforms prompting into an alignment exercise that considers not just the immediate task but also the intended readers, institutional frameworks, and broader social contexts.

3.6 What Goes Wrong: Hallucination, Drift, and Bloat

Even well-intentioned users who write seemingly clear prompts can encounter serious breakdowns in output quality. These failures are often not random but arise from predictable weaknesses in how language models interpret and generate responses. Prompt engineering, therefore, requires not only knowledge of how to make things go right but also insight into how and why things go wrong. This section introduces three of the most common failure modes encountered by users: hallucination, drift, and bloat. Each reflects a distinct type of breakdown that can significantly affect the reliability and usability of an output.

Hallucination refers to the model generating false or fabricated content with apparent confidence. This failure is hazardous in high-stakes domains, such as medicine, law, or academic research, where factual accuracy is non-negotiable. Hallucinations can take many forms: citing non-existent sources, inventing statistics, or producing plausible-sounding but incorrect historical or technical details. They occur because the model is not verifying facts against a grounded database in real time; instead, it is drawing on patterns of language observed during training. Unless explicitly prompted to verify its sources, qualify its certainty, or flag speculation, the model may default to authoritative phrasing even when it is effectively making a guess.

Drift, by contrast, occurs when the model gradually veers off-topic, misinterprets the scope of a task, or begins to introduce irrelevant information. This kind of output degradation is common in long or underspecified prompts. For instance, a request to "explain the ethical risks of AI in healthcare" might yield an answer that begins well but eventually digresses into AI applications in education or finance. Drift reflects the model's tendency to fill space with statistically probable content once it loses clear structural anchors. Without clear limits on scope, length, or format, the model may attempt to be helpful by elaborating far beyond the intended topic.

Bloat refers to the excessive verbosity or redundancy in an AI-generated response. This may take the form of repeated phrases, overly cautious qualifications, or generic filler content that adds length without adding value. While not as hazardous as hallucination, bloat undermines clarity and efficiency. It is especially problematic in contexts where users must quickly scan or repurpose outputs, such as executive summaries, briefings, or system-generated reports. Prompts that fail to specify conciseness, target audience, or intended use leave the model too much room to over-elaborate. The model often defaults to a bloated style when users fail to define the rhetorical task precisely.

These three failure modes, hallucination, drift, and bloat, are symptoms of the same underlying issue: a mismatch between user intent and the model's default generative behavior. Effective prompting seeks to close that gap by making instructions more explicit, structured, and bounded. Techniques such as scaffolding, role definition, fact-checking, and format specification can all help mitigate these failures. For instance, asking the model to cite its sources, summarize in bullet points, or respond "only within the following categories" activates internal constraints that reduce risk.

However, even with careful prompting, some errors remain unavoidable. This situation does not indicate poor prompt design but instead reminds us that language models function as pattern engines rather than knowledge agents. These systems do not "know" what is true or false; they simulate probable answers based on patterns. This fundamental limitation means that users must remain vigilant and critical, not only of what the model says but of how it says it and why. Later chapters extend these safety principles by showing how suppression prompts, disclaimers for advice, and explicit safety constraints help mitigate unacceptable outputs in sensitive or high-risk contexts.

Understanding these failure modes is not merely a technical matter. It is pedagogical. Teaching users how to identify and address hallucination, drift, and bloat prepares them to become more discerning prompt engineers, readers, and writers of AI-generated text. In the next section, we explore how to engage the model not as a one-time responder but as a participant in an iterative dialogue, enabling users to course-correct and refine outputs in real time.

3.7 Reframing and Refinement: Prompting As Dialogue

Prompting is not a one-shot transaction but an evolving conversation between the user and the model. While many users treat AI interaction as a static query–response mechanism, akin to typing a question into a search engine, practical prompt engineering demands a more dynamic mindset. This section introduces a critical principle for improving both output quality and user control: the art of reframing and refinement. It treats prompting as an iterative dialogue rather than a monologue, emphasizing revision, redirection, and layered exploration.

Reframing refers to the process of rewriting a prompt to clarify its goals, narrow its scope, or reposition its intent. This becomes especially important when the initial output from the model is inaccurate, verbose, off-topic, or stylistically unsuitable. Instead of abandoning the task or accepting the flawed output, the skilled user steps back, examines the result, and modifies the original instruction with more precision. For instance, an unsatisfactory answer to "Explain climate change" might be reframed as: "Summarize the primary causes of climate change in 150 words, using plain language suitable for secondary school students." This subtle shift introduces constraints of scope, style, and audience, all of which increase the likelihood of the following output succeeding.

The user refines the output by providing corrective or enhancing instructions, building upon the generated content rather than starting anew. This might include requests such as "Make this more concise," "Add two real-world examples," or "Rewrite in the tone of a policy memo." These directives signal that the user is not merely consuming the AI's response but shaping it, iteratively guiding the model toward alignment with context-specific goals. The refinement process reflects real-world editing practices, where collaborators provide feedback, make revisions, and progressively develop the work through iterative improvement.

This practice reflects a more profound conceptual shift. In traditional human–computer interaction, users issue commands and expect the system to comply. However, with large language models, users are co-steering a generative process that includes ambiguity, interpretation, and stylistic variance. The model is not simply a servant carrying out orders; it is a probabilistic collaborator that offers approximations requiring calibration. Prompting, therefore, becomes a craft of incremental control: guiding the output through cycles of clarification, adaptation, and testing.

The dialogue model also allows for pedagogical scaffolding. By asking the model to "explain your reasoning," "list uncertainties," or "compare your answer to an expert opinion," users can make the AI more transparent and self-aware, not because the model truly understands, but because it can simulate self-explanation. In educational settings, this supports critical thinking. In professional contexts, it adds traceability to decision support. In creative work, it enables users to explore multiple stylistic or conceptual pathways before selecting a final version.

Viewed in this light, prompting becomes an interactive design process rather than a linear transaction. Just as a graphic designer iterates on layouts or a software engineer refactors code, the prompt engineer revises questions and instructions in order to refine the system's behavior. Importantly, this mindset also supports error recovery. When the model produces hallucinated content or drifts off topic, the user can respond not with frustration but with correction: "Only use verifiable sources," or "Exclude information not relevant to EU data policy."

Mastering reframing and refinement means acquiring both strategic patience and analytical agility. It encourages users to move beyond the idea of prompting as a fixed skill and embrace it as an ongoing practice, a way of thinking through problems dialogically, with the AI as a malleable partner. As the next section will demonstrate, even the terminology we use to describe prompts, failures, and formats influences how we learn, teach, and evaluate success in this emerging discipline. A compact worked example in Chapter 4 demonstrates how reframing, chaining, and refinement can be combined in a single multi-step sequence.

3.8 The Language of Prompt Engineering: A Glossary for Everyone

As prompt engineering matures into a professional discipline, it brings with it a new vocabulary. Like any technical field, it relies on specific terms to describe processes, behaviors, and design patterns that would otherwise be ambiguous or misunderstood. However, this emerging language often borrows from computer science, linguistics, and machine learning, making it inaccessible to those without prior training in these fields. This section offers a concise, reader-friendly glossary of key terms, embedded within explanatory sentences to support fluency rather than rote memorization.

The term "prompt" itself refers to the user's input that guides the language model's response. It is not simply a question or instruction, but a designed message with structural and strategic intent. Variants of this concept include "instructional prompting," which issues direct commands, and "role-based prompting," where the user assigns a persona or function to the model, such as "Act as a financial analyst."

Zero-shot prompting represents another core principle, where the model completes tasks based solely on a description without examples, drawing entirely on its trained capacity for generalization. In contrast, "few-shot prompting" includes one or more examples to help the model infer the expected pattern. This distinction is crucial for understanding how models infer task structure without explicit programming.

The context window defines the model's maximum information capacity at any given time, measured in tokens rather than words. When users exceed this limit, the model may discard earlier conversation elements, potentially causing response drift or inaccuracies. Anchoring strategically positions crucial instructions at the prompt's beginning to guarantee their prioritization. Grounding, by contrast, anchors model outputs to specific, verifiable sources rather than relying on general knowledge.

The term "hallucination" has become shorthand for the model's tendency to generate plausible but false content. This term may include fabricated statistics, citations, or even fictitious people. Related to this are "drift," where the model gradually veers off topic, and "bloat," where it provides excessive or redundant information. These failure modes often stem from vague or overlong prompts and require more precise framing to avoid.

"Scaffolding" refers to the technique of breaking complex tasks into stages or layers, enabling the model to reason more clearly and systematically. This technique encompasses both "step-by-step prompting," where the user specifies each stage of a task in a single input, and "turn-based prompting," where outputs are evaluated and refined in sequence. Related to these is "chaining," in which the output of one prompt becomes the input to the next, forming a pipeline of instructions.

Throughout this chapter, scaffolding refers to breaking a task into internal stages, chaining links the output of one prompt to the next, and nested prompting combines multiple instructions inside a single prompt rather than across sequential turns.

Suppression explicitly instructs the model to omit certain content types, such as background explanations or examples, while focus prompting concurrently restricts responses to specific domains, audiences, or length requirements. These techniques are essential for controlling verbosity and ensuring that outputs meet specific constraints.

The glossary of prompt engineering also includes social and ethical terms. “Anthropomorphism” describes the tendency to project human traits, such as intention or understanding, onto the model, a common misconception that undermines effective use. “Alignment ensures model outputs adhere to human values, user intentions, and institutional objectives, while meta-prompting directs the model to evaluate its responses through simulated self-critique or reflective improvement processes.

By weaving these terms into everyday prompting practice, users move from intuition to intentionality. They gain the vocabulary needed not only to describe what works or fails but also to teach, troubleshoot, and iterate. This glossary is not just a reference; it serves as a conceptual bridge that supports the transition from casual use to proficient literacy. In the following chapter, these terms will become increasingly essential as we explore the architecture of complex prompt systems and their application in real-world tasks. A short comparative table appears in Chapter 4 to summarize how these structuring methods differ in purpose and sequencing.

3.9 Summary

This chapter explained the move from informal prompting habits toward intentional prompt literacy. It outlined why LLMs do not behave like search engines or rule-based chatbots and showed how this change requires users to adopt structured instruction rather than conversational phrasing. Core linguistic principles were introduced, including clarity, order, and scope, supported by examples of one-shot, role-based, and instructional prompting.

The chapter also presented practical control techniques for tone, length, and output format and examined three recurring failure modes: hallucination, drift, and bloat. It demonstrated how reframing and iterative refinement strengthen control and improve alignment. A short glossary concluded the chapter by providing shared terminology that prepares readers for more advanced prompting methods in subsequent chapters. The quiz questions at the end of the chapter now test recognition of chaining, nested structuring, and conditional prompts as part of core prompt literacy skills.

3.10 Self-Assessment Quiz

1. What is the primary purpose of prompt literacy in the context of interacting with large language models?

    a) To memorize standard AI outputs and reuse them across tasks

    b) To format responses in machine-readable syntax

    c) To enable clear, intentional, and structured communication with LLMs

    d) To evaluate the internal accuracy of transformer models

2. Which of the following best defines "hallucination" in language model output?

   a) A vivid narrative written in metaphorical language

   b) An output that is too short to meet the prompt's intent

   c) A fabricated or inaccurate response presented as fact

   d) A creative response generated in poetic form

3. Why are role-based prompts considered adequate for shaping AI responses?

   a) They reduce the number of tokens used in generation.

   b) They allow the model to assume a predefined perspective or function.

   c) They turn off other model capabilities unrelated to reasoning.

   d) They allow the user to bypass linguistic ambiguity.

4. What is the term used to describe when a model's output gradually deviates from the intended topic or task?

   a) Compression

   b) Framing

   c) Drift

   d) Reinforcement

5. How can users reduce ambiguity and increase control when designing prompts for LLMs?

   a) By avoiding technical vocabulary and sticking to metaphors

   b) By including more emojis and punctuation marks

   c) By specifying scope, tone, structure, and intended audience

   d) By allowing the model to choose its response format

CHAPTER 4

Advanced Prompting Strategies for Reasoning, Reflection, and Control

As large language models (LLMs) become increasingly embedded in decision-making, education, policy, and professional communication, the design of effective prompts has evolved beyond fundamental command-response interactions. Foundational prompting, such as clarifying roles, setting tone, and defining task format, remains vital. However, many real-world tasks demand more: structured reasoning, deliberation across alternatives, layered self-correction, and procedural thinking. These complexities require a shift in approach: from issuing prompts as simple commands to constructing prompts as architectures, that is, sequential frameworks that guide the model's simulated cognition.

This chapter presents a set of advanced prompting strategies that provide such architectural control. These techniques do not merely elicit answers; they influence how the model reasons, questions, critiques,

H. Tavakoli, *Prompt Engineering for Everyone*,
https://doi.org/10.1007/979-8-8688-2338-1_4

and refines its outputs. By shaping the model's internal processes, these strategies enhance accuracy, interpretability, and relevance, particularly in tasks that involve ambiguity, risk, or creativity. Importantly, they bridge human critical thinking and machine-generated output by imposing structure on the model's probabilistic pattern-matching. From Chain-of-Thought prompting to modular prompt pipelines, the strategies introduced in this chapter constitute the foundational toolkit of the professional prompt engineer. These methods systematically guide outputs while maintaining transparency, ethical oversight, and contextual precision.

A brief note later in the chapter distinguishes among related techniques, including stakeholder, scenario, and reflective prompting, to minimize conceptual overlap.

4.1 Chain-of-Thought Prompting: Structuring Stepwise Reasoning

Chain-of-Thought (CoT) prompting is a foundational strategy in advanced prompt design. It instructs the model to produce a reasoned response by explicitly breaking down a problem into intermediate steps before delivering a final answer. This approach mirrors human methods of analytical thinking, particularly in contexts such as mathematics, diagnosis, legal analysis, and sequential problem-solving. By decomposing a task into parts, CoT prompting renders the model's logic traceable and less prone to error or oversimplification.

The principle underlying CoT prompting is straightforward: simulate structured reasoning through explicit verbal cues. Prompts such as "Let us think step by step," "Break it down logically," or "Explain the reasoning before answering" signal the model to activate latent patterns associated with analytical texts in its training corpus. These include instructional materials, exam solutions, and technical walkthroughs, which frequently model the exact type of breakdown the user seeks to understand. Rather than jumping directly to a potentially superficial conclusion, the model responds with a sequence of operations or justifications, each leading incrementally to the result.

This method is particularly effective in contexts requiring traceability, such as quantitative problem-solving, decision-making with dependencies, or multi-part ethical analysis. Consider the example: "What is 18% of 75?" A standard prompt may return "13.5" without elaboration. A CoT prompt, "What is 18% of 75? Let us think step by step," typically triggers a more structured response:

(1) Convert 18% to 0.18; (2) Multiply 0.18 × 75 = 13.5; (3) Final Answer: 13.5.

Researchers can extend this pattern to more complex domains, such as causal inference or policy assessment, where the model articulates its logic before proposing a solution.

Chain-of-Thought prompting enhances model accountability by exposing its reasoning steps, giving users opportunities to verify, critique, and correct the process. This transparency helps prevent undue reliance on seemingly confident but potentially erroneous outputs, a known limitation of large language models. Moreover, it supports educational objectives by enabling learners to observe and internalize structured reasoning strategies. In professional settings, it enhances the auditability and defensibility of AI-generated recommendations. Researchers can adapt Chain-of-Thought (CoT) prompting to various rhetorical or disciplinary contexts by using formulations such as "Walk through your reasoning," "Take it one part at a time," or "Talk me through the logic." Combining these cues with formatting techniques such as bullet points, indentation, or numbered steps improves clarity. However, practitioners should use CoT judiciously, as simple, open-ended, or creative tasks may become unnecessarily verbose or constrained when processed through this method.

In sum, Chain-of-Thought prompting transforms LLMs from answer generators into process models. It enables human users to guide, monitor, and refine the model's simulated cognition, ensuring that reasoning is not merely present but visible and open to scrutiny. As such, it represents a core strategy in the emerging discipline of prompt architecture, where the prompt is not simply a query, but a thinking design.

4.2 Tree-of-Thought Prompting: Divergent Reasoning and Strategic Evaluation

Tree-of-Thought (ToT) prompting extends the principle of stepwise reasoning by introducing multiple, parallel reasoning pathways. Rather than directing the model to follow a single line of logic from question to conclusion, ToT prompting invites it to generate several alternative solutions, assess their relative merits, and converge on the most suitable outcome. This approach mirrors deliberative reasoning processes that are standard in planning, strategy, and ethical decision-making contexts, where decision-makers must evaluate multiple viable options against competing criteria, rather than identifying a single correct solution.

Whereas Chain-of-Thought prompting relies on linear decomposition, Tree-of-Thought prompting operates by branching. Each branch represents a discrete hypothesis, proposal, or interpretive stance. The model is encouraged to explore these branches separately, articulating the reasoning behind each and identifying the implications or trade-offs associated with them. Only after examining the complete set of alternatives does the model arrive at a reasoned judgment or final recommendation. This recursive structure produces richer and more balanced outputs, especially in domains where ambiguity, conflict, or uncertainty is unavoidable.

Consider the prompt: "What are three viable strategies for reducing carbon emissions in urban areas? For each, list key benefits and drawbacks. Then choose the most effective one and justify your choice." Tree-of-Thought prompting enables models to generate multiple policy responses, such as expanding public transit, imposing carbon pricing, and subsidizing green retrofits, before systematically evaluating each option against criteria like feasibility, cost, equity, and environmental impact. The model then synthesizes this analysis into a final position, replicating the deliberative processes found in government reports, academic debates, and multidisciplinary panels while diversifying its outputs.

ToT prompting proves particularly valuable in fields such as public policy, systems design, product strategy, and risk management, where stakeholders must navigate trade-offs. By articulating multiple possibilities and inviting structured comparison, the technique helps surface blind spots and mitigates the risk of tunnel vision. It also aligns with practices of inclusive design and collaborative evaluation, supporting outputs that are sensitive to pluralistic values and institutional constraints.

The method is also adaptable. Users may instruct the model to produce a fixed number of branches or vary the number according to complexity. Evaluation criteria may be supplied in advance, e.g., "Rank options by cost-effectiveness and implementation time," or elicited by the model itself. More sophisticated applications assign distinct roles to different branches, such as a technical expert, community advocate, or policymaker, enabling simulated stakeholder consultations that mirror the diversity of real-world decision-making contexts.

However, the effectiveness of Tree-of-Thought prompting depends on well-structured instructions. Without clear guidance, the model may collapse divergent paths into a superficial summary or fail to distinguish meaningfully between options. Excessive complexity risks creating token inefficiency or cognitive overload, particularly when practitioners introduce too many branches or criteria without establishing a proper hierarchy or priority.

When implemented carefully, Tree-of-Thought prompting transforms the model from a reactive respondent into a deliberative agent. This approach produces more considered and defensible outputs, particularly valuable when stakeholders need to explain, justify, or debate the resulting decisions. As large language models increasingly participate in strategic, professional, and policy-oriented workflows, this technique provides a blueprint for pluralistic reasoning at scale.

4.3 Self-Ask Prompting: Internal Questioning for Decomposable Tasks

Self-Ask prompting is a strategy that encourages the language model to engage in internal questioning before arriving at a final answer. Unlike methods that follow a predefined reasoning structure (as in Chain-of-Thought or Tree-of-Thought prompting), Self-Ask relies on the model's ability to identify and answer a sequence of intermediate sub-questions that it generates independently. Self-Ask prompting simulates procedural self-inquiry by making the model pause and determine what information it requires before responding. This method enhances transparency, traceability, and accuracy, particularly for tasks demanding multi-layered factual inference or complex calculations.

The pedagogical analogue for Self-Ask prompting is the Socratic method: the model interrogates the problem space as a thoughtful learner might, asking clarifying questions that progressively narrow ambiguity or isolate needed variables. For example, if given the query, "How many people in China live in urban areas?", a typical Self-Ask response might unfold as follows:

(1) What proportion of China's population resides in urban regions? → 65%.

(2) What is China's total population? → 1.4 billion.

(3) What is 65% of 1.4 billion? → 910 million.

Only then does the model deliver the final answer. Each sub-question reveals a discrete step in the reasoning chain, allowing both the user and the model to verify correctness along the way.

This technique is beneficial in contexts where the input prompt is underspecified or decomposable, tasks that benefit from being broken into constituent steps, but where the nature of those steps is not immediately apparent. Self-Ask prompting differs from Chain-of-Thought prompting in that it requires the model to identify knowledge gaps, simulate missing information, and fill them through staged elaboration, rather than following user-predefined logic. This approach fosters generative and adaptive machine reasoning.

From an engineering perspective, Self-Ask prompting increases the interpretability of model outputs by surfacing latent assumptions. In data analysis, the technique can expose flawed premises or highlight missing variables. In technical writing and scientific interpretation, it ensures that underlying calculations or deductions are visible to the user. In legal or ethical evaluation, Self-Ask prompting helps reveal the implicit questions that drive judgment.

There are several best practices for applying this technique effectively. Prompts should explicitly encourage question generation, e.g., "Begin by asking what you need to find out," and may optionally guide the format, such as by requesting bullet points or numbered stages. Limiting the scope of inquiry (e.g., no more than three sub-questions) can improve token efficiency and avoid unnecessary elaboration. To ensure quality, users should inspect each question–answer pair for logic and factual consistency, especially in high-stakes applications.

Nevertheless, the method has its limitations. Because the model must autonomously select which questions to ask, there is a risk of mis-framing the problem or overlooking key variables. In cases where precision or domain-specific knowledge is critical, user oversight remains essential.

Furthermore, the additional reasoning steps consume more tokens than single-response strategies and may be unsuitable for time-constrained or low-bandwidth contexts.

Despite these constraints, Self-Ask prompting serves as a powerful method for inducing structured, self-directed reasoning in language models. It encourages the model not only to generate content but to simulate curiosity, build logical scaffolds, and foreground the internal questions that underpin its conclusions. This approach particularly benefits pedagogical, analytical, and exploratory tasks that prioritize transparency, procedural logic, and iterative learning.

4.4 Reflection and Self-Critique Prompting: Simulated Revision and Quality Control

Reflection and self-criticism prompting are advanced strategies that instruct a language model to review, evaluate, and revise its output before finalizing a response. Rather than treating the model's first-generation output as complete, these methods frame the generation process as iterative, requiring internal feedback loops that mimic expert editorial practice. This approach improves coherence, reduces hallucinations, and enables the model to simulate a form of metacognitive reasoning, which is beneficial in tasks where rigor, balance, or accountability is essential.

At its core, reflection prompting introduces a two-stage process: the model first generates an initial answer, then engages in a critical self-assessment of that output. The prompt typically includes instructions such as "Now critique your response and revise it for clarity and completeness" or "Evaluate the tone and accuracy of your answer and improve it." This simulation of revision triggers the model to identify possible omissions, inconsistencies, or stylistic issues before producing a revised version. The output, therefore, reflects a second layer of processing that is more deliberate and refined.

The analogy here is that of an experienced editor reviewing a colleague's draft, not only for errors but for audience alignment, logical structure, and argumentative strength. While the model lacks awareness or intent, it can simulate evaluative heuristics learned from revision-rich textual data, particularly in academic, technical, and journalistic domains. These patterns allow it to detect common flaws and articulate reasoning about potential improvements, even in the absence of ground truth.

Reflection prompting is particularly useful in fields that demand a high standard of precision or where outputs are subject to professional review, such as academic writing, policy briefs, legal summaries, technical documentation, and formal correspondence. It is equally valuable in educational settings, where it models revision practices for students and offers insight into the decision-making processes behind writing. In each case, the technique elevates the language model from a first-draft generator to a more reliable co-authoring tool.

Researchers can scaffold reflection prompting using several formats, with the *"Answer → Reflect → Revise"* pattern representing one standard structure. Other builds in a checklist: for example, "After writing your answer, check whether it (a) addresses all aspects of the question, (b) maintains a consistent tone, and (c) uses appropriate terminology. Then revise based on your findings." In advanced implementations, the method simulates peer review by instructing the model to adopt a domain expert's perspective, first critiquing the initial output before producing a revised version. These various formats correspond to distinct professional standards and expectations.

Researchers must employ reflection prompting judiciously, as the model might critique superficial features while overlooking deeper conceptual or factual errors, particularly those beyond its training boundaries. Moreover, the revision process can sometimes result in overcorrection, simplifying, generalizing, or hedging to avoid risk. This process can dilute argumentative strength or compromise the specificity of the original answer. Human oversight remains essential, particularly in high-stakes or specialized applications.

Nonetheless, the strategic integration of reflection and self-criticism can significantly enhance prompt outcomes. By embedding a second layer of reasoning into the generation process, users can elicit responses that are not only more polished but also more aligned with task goals, stylistic requirements, and domain conventions. These techniques foster a culture of iterative improvement in human–AI collaboration, one that mirrors best practices in writing, design, and research.

4.5 Multi-agent Prompting: Simulating Dialogue and Deliberation

Multi-agent prompting is a prompting strategy that directs a language model to simulate dialogue among multiple distinct agents, each representing a specific role, perspective, or area of expertise. The model engages in structured conversations, debates, critiques, or negotiations among simulated participants, rather than producing a singular output. This approach mirrors real-world decision-making processes, where collaborators weigh and synthesize multiple viewpoints before reaching conclusions.

At the core of this technique lies the premise that internal diversity fosters external quality. This multi-agent approach guides the model to surface tensions, explore trade-offs, and present alternative interpretations that single-voice prompts might not reveal. A basic multi-agent prompt might read: "Consultant A supports Strategy X for reducing carbon emissions. Consultant B is skeptical and raises ethical concerns. Simulate their discussion and provide a balanced recommendation." The model then generates an exchange in which these viewpoints are articulated, challenged, and integrated.

This technique is particularly well-suited to complex, ambiguous, or value-laden domains, such as public policy, ethics, legal analysis, or interdisciplinary research, where no single perspective can provide a complete or definitive answer. By prompting the model to inhabit contrasting roles, users gain access to richer output that mirrors real-world intellectual and professional diversity.

Researchers may implement multi-agent prompting with two or more participants, assigning specific instructions to each agent. For instance, one may focus on cost-efficiency, another on social justice, and a third on regulatory feasibility. By assigning distinct cognitive or ethical commitments to each role, the user creates a framework in which reasoning becomes comparative, layered, and iterative. Professional practice uses this approach to simulate committee discussions, stakeholder consultations, or expert panel scenarios that require decision-makers to reach rigorous conclusions while balancing competing priorities.

A key advantage of this approach is the mitigation of model overconfidence and uncritical synthesis. When roles are clearly defined and in tension, the model is less likely to converge prematurely on a single, unsupported answer. It becomes structurally committed to elaborating multiple positions and justifying any synthesis that emerges. In turn, this encourages interpretability and transparency, qualities that are vital in decision support systems, legal frameworks, and AI-aided governance.

Nevertheless, effective multi-agent prompting requires precision in role specification. Without clear differentiation, the model may conflate perspectives or revert to a neutral, generic tone. It is also important to define the interaction structure, whether the dialogue should conclude with a consensus, end in disagreement, or present parallel positions for the user to evaluate. Poorly scoped prompts may lead to incoherence or redundancy, diminishing the value of the exchange.

This technique also raises ethical considerations. When simulating roles tied to cultural identity, political ideology, or lived experience, users must design prompts with sensitivity and avoid stereotypes. The roles should be grounded in institutional or professional archetypes, such as "environmental economist," "public health officer," or "cybersecurity auditor," rather than social caricatures.

In sum, multi-agent prompting transforms the model from a monologic generator into a polyphonic simulator. It enables users to orchestrate complex internal conversations, anticipate counterarguments, and generate recommendations that are both more robust and more representative of diverse stakeholder positions. As language models continue to support critical decisions in education, policy, healthcare, and beyond, this technique offers a powerful means of embedding deliberation into the generation process, making AI output not just intelligent but also dialogically informed.

4.6 Iterative Refinement Prompting: Improving Through Successive Steps

Iterative refinement prompting is a structured technique that guides a language model through multiple stages of controlled revision. Rather than expecting a perfect output from a single prompt, this method breaks down content development into successive improvement cycles, each targeting a specific dimension of quality such as structure, clarity, tone, or completeness. This approach aligns with the editorial principle that professional outputs typically emerge from successive drafting, not instantaneous generation.

Iterative prompting follows a simple yet powerful logic: prompting a model to refine an existing output rather than start from scratch enables it to focus more narrowly on one dimension of improvement. For example, a user might first instruct: "Draft a 100-word summary of this article." Once the model generates a preliminary version, the user could then prompt: "Now revise this for clarity and eliminate repetition." A subsequent prompt might ask: "Polish the tone for an academic audience." Each instruction incrementally shapes the output into a more effective and purpose-specific form.

The process replicates real-world workflows in publishing, technical documentation, UX writing, and strategic communication, where professionals typically develop documents through phased reviews and collaborative iterations. Rather than expecting the language model to produce a finalized artifact on its first attempt, iterative refinement enables the user to serve as a supervising editor, evaluating the model's response at each stage, issuing targeted revision instructions, and redirecting when necessary.

Researchers can operationalize the method in two primary forms: fixed-step and adaptive refinement. Fixed-step sequences predefine each stage of revision in advance, as in (1) generate draft, (2) fix grammar, (3) improve tone, and (4) add examples. This plan is ideal for standardized workflows or templated outputs. In contrast, adaptive refinement responds dynamically to quality issues in the model's initial response. The user diagnoses weaknesses and prompts the model to fix them sequentially. This approach is better suited for bespoke outputs or exploratory writing.

A core benefit of iterative refinement is error isolation. When the user targets only one dimension at a time, such as verbosity, tone, or factual completeness, problems are easier to detect and correct. This method also reduces one of the central risks in prompt engineering: compound errors that occur when practitioners bundle multiple tasks into a single, overloaded instruction. By distributing these tasks across stages, the model performs with greater focus and accuracy.

Moreover, iterative prompting offers a mechanism for quality assurance in sensitive contexts. In professional and high-stakes domains, such as legal analysis, grant writing, or medical summarization, each revision phase becomes an opportunity for human oversight. This human-in-the-loop framework allows the user to intervene, approve, or redirect the model's trajectory without losing the overall structure of the work. The result is an AI-assisted authoring process that strikes a balance between efficiency and accountability.

Nonetheless, iterative refinement has trade-offs. Because each phase consumes tokens, the approach increases computational cost. Additionally, if revision instructions are vague, e.g., "Make it better," the model may misinterpret the goal or degrade previous improvements. Therefore, clear and constrained revision prompts are essential. Phrases such as "Simplify sentence structure while preserving meaning" or "Add a real-world example, no more than two sentences" ensure focused progress without introducing drift or bloat.

In summary, iterative refinement prompting transforms the act of generation into a collaborative, layered process. It aligns AI output with the cognitive rhythm of human revision: draft, diagnose, revise, and polish. Rather than a tool for one-shot content production, the model becomes a co-editor, responsive, adaptable, and capable of producing professional-grade outputs through disciplined iteration. As language models become increasingly integrated into institutional workflows, this technique will be crucial for ensuring quality, reliability, and alignment with human intent.

4.7 Creative Prompting: Structured Divergence

Creative prompting is a strategy that invites language models to generate novel, stylistically distinct, or imaginative outputs while remaining grounded in purposeful constraints. Unlike prompts that seek factual precision or procedural logic, this technique prioritizes expressive variation, rhetorical flexibility, and idea generation. Its value lies in balancing freedom with form, enabling divergent thinking without sacrificing clarity, tone, or contextual relevance.

This method draws on a well-established principle in creativity research: constraints foster invention. Rather than instructing the model to "be creative," effective prompts specify parameters such as stylistic range, output tone, audience sensitivity, or rhetorical stance. The framework defines boundaries while encouraging variation, empowering the model to explore multiple interpretive possibilities that respond to the same underlying input but present it in stylistically or structurally different ways.

For example, a creative prompt might request three campaign slogans for a health awareness initiative, one formal, one poetic, and one satirical. Alternatively, a user might instruct the model to retell a historical event in the voice of a fictional character, translate an economic concept into a metaphor involving bees, or generate three explanations of a scientific process suitable for children, teenagers, and professionals, respectively. In each case, the goal is not just linguistic fluency, but variation across affective tone, narrative framing, or conceptual metaphor.

Creative prompting is particularly effective in fields such as brand messaging, public education, speculative design, content marketing, and interdisciplinary research communication. In these domains, outputs must not only inform but engage diverse audiences, sometimes through emotional resonance, humor, surprise, or aesthetic framing. Relying solely on informative prompts risks flattening communication into generic or overly literal outputs. By contrast, structured divergence enables the model to contribute to ideation, mood setting, and even cultural signaling.

Importantly, creativity in prompting does not imply randomness or disregard for the audience and context. On the contrary, it requires careful guidance. Overly open-ended prompts may result in incoherence, offensive outputs, or stylistic mismatch. Professional use of creative prompting involves precise specification of tone, structure, intended function, and boundaries, especially when content must align with organizational values or public sensibilities.

As with other advanced strategies, creative prompting can be modularized and iterated. Users may request multiple outputs in parallel ("Generate five metaphors") and then refine the most promising one using reflection or focus techniques. In collaborative settings, this process enables the collective selection and refinement of model-generated ideas, thereby anchoring AI-driven creativity within structured review cycles.

Ultimately, creative prompting transforms language models from data processing tools into co-designers of meaning. It equips users with a method for stimulating linguistic novelty while maintaining control, intent, and responsibility. As LLMs become embedded in cultural, educational, and communicative systems, the capacity to guide them toward expressive and adaptive outputs will become as essential as factual correctness or logical coherence.

A short clarification later in the chapter also distinguishes creative prompting from reflective and stakeholder techniques to avoid conceptual overlap.

4.8 Prompt Pipelines: Building Modular Workflows

Prompt pipelines are an advanced strategy that decomposes complex tasks into a structured sequence of interdependent prompts. Instead of relying on a single, monolithic instruction, the user guides the language model through multiple coordinated stages, with each prompt fulfilling a distinct role. This method reflects professional workflows commonly found in software development, policy design, publishing, and education domains, where clarity and process integrity depend on modular execution.

The core principle of prompt pipelines is serial composition. Each stage of the pipeline produces output that serves as input for the next, enabling compound reasoning, progressive refinement, or role-based transformation. This structure allows users to isolate specific subtasks, such as extraction, classification, elaboration, or translation, and design prompts optimized for each. The result is greater precision, easier debugging, and more transparent control over the overall generation process.

Consider a policy analysis task. A single prompt may struggle to strike a balance between depth, clarity, and conciseness. Instead, a pipeline might begin with a summarization prompt that extracts key issues from a report, followed by a classification prompt that organizes those issues by policy domain, and conclude with a recommendation prompt tailored to the needs of a target audience. Each prompt in the sequence is simpler than the combined task, but together they produce a more coherent, context-aware result.

Prompt pipelines are especially valuable in high-stakes or collaborative settings. Because each prompt is functionally discrete, teams can assign different stages to different users or departments. For example, a communications team might refine the tone of a technical summary generated by a research analyst upstream. This division of labor also enables asynchronous workflows, version control, and quality assurance at multiple checkpoints.

Pipelines also support experimentation and reuse. Because prompts are modular, users can substitute, iterate, or adjust one part without having to rewrite the entire sequence. This flexibility is beneficial when testing variations in tone, structure, or reasoning approach. For instance, a user might compare the effect of changing only the second-stage prompt, switching from a neutral tone to a persuasive one while keeping all other elements constant.

However, researchers must apply discipline when designing effective pipelines. They should carefully scope each prompt to avoid redundancy or token overflow, clearly articulate transitions between stages, and maintain consistent output formatting to support chaining. In cases where errors occur, users can pinpoint failure points to individual prompts rather than attributing issues to the entire workflow, a significant advantage in professional environments that demand transparency and auditability.

Prompt pipelines are not merely a workaround for model limitations; they are a crucial component of model development. They represent a shift in how language models are deployed, from static responders to dynamic components in distributed problem-solving systems. Just as software developers modularize code to improve clarity and reuse, prompt engineers can modularize language interactions to enhance flexibility, maintainability, and alignment with institutional goals.

By learning to construct and iterate on prompt pipelines, users move beyond isolated outputs toward building sustainable, adaptive interfaces between human intent and machine response. This skill marks an important threshold in prompt literacy: the transition from writing prompts to engineering structured, reliable dialogue systems.

A brief transition note here signals how modular pipelines differ from stakeholder and reflective strategies without repeating earlier conceptual material.

4.8.1 Strategy Selector: Matching Prompting Method to Task Demands

The choice of strategy depends on two variables: task complexity and risk level. Low-complexity, low-risk tasks (e.g., simple calculations, everyday summaries) benefit from step-by-step reasoning through Chain-of-Thought prompting, which offers transparency without unnecessary overhead. As complexity increases but risk remains modest (e.g., creative ideation, exploratory analysis), Self-Ask prompting or creative prompting enables divergence and novelty while tolerating minor inaccuracies.

High-complexity, moderate-risk contexts (e.g., policy options, technical design debates) require Tree-of-Thought prompting or multi-agent prompting, as these methods facilitate the surfacing of alternatives and support structured evaluation. Finally, high-complexity, high-risk tasks (e.g., medical decision support, legal drafting, sensitive governance) demand reflection, iterative refinement, and prompt pipelines. These strategies maximize factuality, consistency, and oversight by embedding review and correction loops into the workflow.

The resulting framework may be visualized as a two-dimensional grid:

- **Simple** × **low-risk** → Chain-of-Thought
- **Simple** × **high-risk** → Reflection/iterative refinement
- **Complex** × **low-risk** → Self-Ask/creative prompting
- **Complex** × **high-risk** → Tree-of-Thought/multi-agent/pipelines

This matrix highlights not only how to apply each technique but also when to prioritize deliberation, critique, or modular control in practice.

A brief comparative note later in the chapter clarifies how stakeholder prompts differ from scenario prompts in applied settings.

4.8.2 Worked Comparative Example: Applying Multiple Strategies to the Same Task

To illustrate the distinct affordances of advanced prompting techniques, consider the task: *"Provide a policy recommendation for reducing traffic congestion in a large metropolitan area."* The following comparison applies four strategies, Chain-of-Thought, Tree-of-Thought, Reflection, and Multi-agent prompting, to the same task, highlighting their respective strengths and trade-offs.

4.8.3 Chain-of-Thought Prompting

Using a step-by-step instruction such as *"Explain your reasoning step by step before recommending a policy,"* the model decomposes the problem: identifying causes (car dependency, limited transit, poor cycling infrastructure), then examining interventions, and finally recommending expanded public transport. This approach yields moderate latency and token usage, as reasoning is linear; however, factual consistency is high, as each step is traceable. The limitation is a reduced breadth; alternative solutions may be overlooked because the focus remains on a single line of reasoning.

4.8.4 Tree-of-Thought Prompting

When prompted to *"Generate three possible strategies, outline benefits and drawbacks of each, then recommend the best,"* the model develops parallel branches: congestion pricing, public transit expansion, and telecommuting incentives. Each is evaluated for feasibility, equity, and cost. Latency and token use are higher than Chain-of-Thought, but the factual coverage is broader. The output is richer, surfacing trade-offs that linear reasoning alone might conceal. This makes Tree-of-Thought especially suitable when multiple competing options require deliberate evaluation.

4.8.5 Reflection Prompting

Here, the model is first asked to propose a solution, then to critique and revise it. For instance, the initial answer may emphasize congestion pricing. The follow-up instruction *"Now review your answer: does it balance economic, social, and environmental factors? Revise accordingly"* prompts self-correction. The revised output often integrates additional measures, such as subsidies for low-income commuters or parallel investment in bus infrastructure. Latency and tokens increase due to the double-pass structure, but factuality and internal coherence improve. This strategy is particularly effective when reliability and balance are paramount.

4.8.6 Multi-agent Prompting

Finally, when the model is asked to simulate dialogue, for example, between a transport economist, an environmental activist, and a city planner, the output becomes a deliberative exchange. Each agent raises discipline-specific concerns: efficiency, sustainability, and urban design. The resulting synthesis, while the most token-intensive and slowest to generate, reflects plural perspectives and surfaces hidden tensions, such as social equity vs. fiscal feasibility. This strategy maximizes interpretability and inclusiveness, though at a higher computational cost.

4.8.7 Comparative Summary

Across these four strategies, a consistent pattern emerges. Chain-of-Thought is efficient and reliable but narrow; Tree-of-Thought balances breadth with structured comparison; Reflection enhances accuracy through revision cycles; and Multi-agent prompting provides the wealthiest, most deliberative outputs at the cost of latency and tokens. Factuality is generally strong across all methods, though Reflection and Tree-of-Thought provide superior safeguards against one-dimensional or incomplete reasoning.

This comparison demonstrates that no single strategy is universally optimal. Instead, the choice depends on the intersection of task complexity and risk level, as outlined in the Strategy Selector matrix. For low-risk, straightforward tasks, Chain-of-Thought may suffice; for high-stakes policy decisions, combining Tree-of-Thought with Reflection or Multi-agent prompting ensures greater robustness.

4.9 Conclusion

4.9.1 A Toolbox for Prompt Strategists

The prompting strategies introduced in this chapter extend the practice of prompt engineering beyond direct instruction and into the realm of structured reasoning, iterative control, and expressive design. Each technique contributes to a broader framework in which prompting is not merely a matter of issuing commands, but of architecting logic, reflection, and intentionality into the interaction between the user and the model.

Chain-of-Thought prompting brings traceability and structure to reasoning tasks, encouraging transparent step-by-step logic. Tree-of-Thought and Self-Ask techniques support divergent analysis and deliberate question decomposition, enabling the model to navigate complex and ambiguous scenarios with greater deliberation and insight. Reflection and iterative refinement elevate the output quality by introducing critical feedback loops, guiding the model through processes of evaluation and revision that parallel expert workflows.

Multi-agent prompting simulates diverse perspectives, turning monologic responses into dialogic exchanges that surface assumptions, trade-offs, and dissent. Creative prompting, in turn, unlocks the model's capacity for stylistic variation, metaphorical expression, and imaginative framing, all within the boundaries of clearly defined communicative objectives. Prompt pipelines introduce modularity and coherence across multi-phase tasks, ensuring that each step contributes meaningfully to the whole. Focus prompting closes the loop by enforcing constraints on scope, tone, or length, sharpening the model's output and aligning it with institutional needs.

Together, these techniques provide a practical and principled foundation for designing intelligent human–AI interaction. They do more than elicit content; they operationalize reasoning, simulate critical engagement, and embody a structured partnership between human intent and machine-generated language. In applying these strategies, the prompt engineer assumes a new role: not only instructing the model but shaping its simulated cognitive process, turning an input/output exchange into a form of design thinking.

Mastery of these advanced prompting techniques is essential for professionals who rely on language models to produce not only fluent but also defensible, structured, and ethically sound outputs. As language models become increasingly integrated into decision-making, education, policy, and knowledge production, prompt engineering must evolve in parallel, transitioning from an intuitive approach to a methodical one and from experimentation to a disciplined process.

4.10 Self-Assessment Quiz

1. What distinguishes Tree-of-Thought prompting from Chain-of-Thought prompting in terms of model behavior?

    a) It focuses on generating poetic responses using stylistic variation.

    b) It requires the model to follow a strict, single reasoning path.

    c) It enables the model to explore multiple solution paths before reaching a conclusion.

    d) It simulates self-critique and revision within one coherent output.

2. Which prompting strategy is most appropriate when a user wants the model to revise a previously generated answer after evaluating its completeness and tone?

    a) Focus prompting

    b) Reflection prompting

    c) Self-Ask prompting

    d) Tree-of-Thought prompting

3. In what scenario is multi-agent prompting most effective?

    a) When a prompt requires purely factual summarization

    b) When a user needs the model to mimic Socratic-style inquiry

    c) When conflicting viewpoints must be simulated to reach a balanced judgment

    d) When generating multilingual outputs across different cultural registers

4. What key benefits do prompt pipelines provide when handling complex workflows?

    a) They reduce the need for user oversight by allowing end-to-end automation.

    b) They allow each prompt stage to operate independently, enabling modular refinement.

    c) They guarantee factual accuracy by querying external sources in each step.

    d) They shorten response time by compressing tasks into a single instruction.

5. Which of the following best characterizes the purpose of focus prompting?

   a) To help the model brainstorm freely without boundaries

   b) To impose structure, reduce token waste, and ensure task alignment

   c) To allow the model to generate role-based dialogues with internal critique

   d) To train the model on user-specific vocabulary through repetition

CHAPTER 5

Precision Prompting: Controlling AI with Granular Instructions

Precision prompting refers to the intentional control of how a language model interprets and responds to a user's input. It is not simply about being detailed; it is about being deliberately structured. In practical terms, precision distinguishes between a general inquiry and a defined task; between a loosely guided conversation and a purposeful design. As prompting becomes a core interface for applied AI, the ability to construct clear, bounded, and context-aware instructions becomes essential for reliability, safety, and ethical alignment.

The following sections briefly distinguish structure from output format and focus from constraints to avoid conceptual overlap.

This chapter builds on the foundational literacy established in the earlier chapters, particularly the ideas introduced in Chapter 3 on clarity, scope, and structure and Chapter 4 on prompting for reasoning and control. Where Chapter 4 explored the creative and cognitive dimensions of prompt architecture, the present chapter shifts focus toward granular control: how to use language to constrain, guide, and verify AI behavior

H. Tavakoli, *Prompt Engineering for Everyone*,
https://doi.org/10.1007/979-8-8688-2338-1_5

at a professional standard. This plan includes specifying tone, format, length, target audience, output style, and task decomposition. In doing so, precision prompting becomes both a linguistic and ethical responsibility, particularly in domains where ambiguity can lead to misinformation, inefficiency, or harm.

Positioned just before Chapter 6 introduces the CASTROFF framework, this chapter establishes the conceptual and practical foundations for developing professional-grade prompting standards. It invites the reader to move beyond experimentation and toward intentionality. By examining scaffolding, prompt layering, templates, and real-world patterns, we begin to see prompting not merely as interaction but as instruction, designed for reproducibility, clarity, and equity.

For non-technical readers, precision prompting offers a way to confidently collaborate with AI systems without needing to write code. For technical audiences, it introduces a language-level toolkit that complements algorithmic control. In both cases, the chapter demonstrates that precision is not about verbosity, nor is it a stylistic preference. It is a necessary condition for effective, fair, and auditable human–AI interaction.

5.1 What Does Precision Mean in Prompting?

Precision is a foundational principle of professional prompt engineering. It marks the distinction between casual user interaction and intentional, structured engagement. A precise prompt allows a language model to produce outputs that are specific, goal-aligned, and contextually relevant, rather than generic or overly broad.

In practice, precision requires supplying the optimal amount of information to steer the model's response effectively. This typically involves defining the task's scope, identifying the target audience, establishing relevant constraints, and specifying the preferred output format and tone. For instance, asking for "something healthy" is vague and leaves much open to interpretation. In contrast, a prompt such as "Create a vegan lunch under 500 calories using only ingredients available in a typical UK supermarket, with a preparation time of under 15 minutes" provides a clear and bounded instruction. The difference lies not in length but in informational density and structural clarity.

Constraints can operate at different levels of strength. "At the soft end, users may set a general preference, such as use plain language." At a medium level, they may define limits on word count or style, while hard constraints specify exact structure or format, such as requiring a JSON output with predefined fields. This constraint ladder helps practitioners determine how firmly to guide the model based on the context.

Importantly, precision is not about verbosity. A concise but well-scoped prompt often outperforms a long, rambling one. Precision elevates prompting from open-ended conversation to structured design, particularly in institutional, regulated, or high-stakes contexts where reliability is critical.

5.2 Precision Scaffolding: Breaking Down Tasks

Precision scaffolding refers to the deliberate structuring of a prompt into clear, manageable components that guide an AI system through a complex task. Just as educators scaffold student learning by breaking down difficult material into sequential steps, prompt engineers use scaffolding to ensure that each part of an AI-generated response is purposeful, coherent, and aligned with the overall goal.

At its core, scaffolding transforms a single abstract request into a layered instruction set. This approach is beneficial when the intended output requires synthesis, evaluation, or structured formatting. Instead of relying on the model to infer how to proceed, scaffolding defines a path, segmenting the task into logical stages. This approach not only improves the quality of the AI's response but also enhances its interpretability and reliability in professional settings.

The scaffolding framework comprises two primary modes. The first mode, step-by-step prompting, outlines the entire task within a single, multi-part prompt, which suits well-defined, repeatable workflows particularly well. For example, a research assistant prompt might request: (1) a list of relevant sources, (2) a summary of each, (3) a comparative analysis in a specific format, and (4) a conclusion tailored to a defined audience.

The second mode, turn-based prompting, executes each stage sequentially and incorporates human review or feedback between steps. This method is especially effective for high-stakes, creative, or evolving tasks. By allowing iteration and refinement at each stage, turn-based prompting introduces an element of co-authorship and control, enabling human oversight while maintaining efficiency.

Consider the following scaffolded sequence designed for generating a policy briefing:

1. Identify and list three recent peer-reviewed articles on the topic.
2. Summarize each article's main findings in no more than 100 words.
3. Present a comparative analysis of the findings using a simple matrix format.
4. Conclude with a summary suitable for a non-expert policymaker, using a formal tone.

Here, scaffolding does more than organize the prompt; it creates checkpoints for transparency, traceability, and quality assurance. In regulated environments or collaborative workflows, such scaffolding helps mitigate risk by clarifying expectations and documenting decision pathways.

More broadly, scaffolding serves as a foundation for auditability in AI interactions. Clear instructions generate outputs that help users track the reasoning behind conclusions and verify their alignment with the intended purpose. This characteristic establishes precision scaffolding as both a methodological tool and an ethical professional practice.

5.3 Prompt Layers: Controlling Depth and Detail

Prompt layering is a method of controlling the internal structure of a prompt to guide the model not only in what to say but also in how to say it. By layering instructions related to content scope, knowledge depth, linguistic tone, response length, and formatting, the prompt engineer creates a multidimensional framework within which the AI operates. This technique is essential when outputs need to meet professional or institutional standards, where the precision of presentation is just as important as the accuracy of the information.

In practical terms, prompt layering allows the user to calibrate the model's output along several axes. One layer might define the required level of depth, specifying whether the explanation should be suitable for a lay audience or assume specialist knowledge. Another might constrain the response length, such as by requesting a summary in 150 words or less or a three-paragraph report. Layering may also determine the desired focus, instructing the model to address a specific aspect, such as causes of a problem without offering solutions, or historical context without policy recommendations. Further, the prompt can control the degree of detail by requesting the inclusion of supporting elements such as statistics, analogies, or real-world examples. The user can also specify the structural form of the output, such as a narrative, list, comparison table, or formally formatted letter. Finally, tone and style are also subject to control, with prompts indicating the need for plain language, a formal tone, a persuasive style, or a domain-specific register.

For example, a prompt might ask the model to write a plain English executive summary explaining how electric vehicles reduce emissions. To ensure clarity and control, the prompt may further require a limit of 150 words, the inclusion of one analogy and one verifiable statistic, and a focus restricted to environmental benefits rather than broader policy or technological issues. Through this layered construction, the user guides the model toward a bounded output that aligns with the communicative goal.

Layering is particularly valuable in contexts where misinterpretation or inconsistency can undermine the usefulness of AI-generated content. In fields such as law, education, healthcare, and policy, the failure to match tone, format, or level of detail to the audience can result in confusion, non-compliance, or reputational harm. Prompt layering, therefore, serves not only as a design strategy but as a safeguard, making outputs more predictable, auditable, and professionally appropriate. In later chapters, CASTROFF is introduced in the same sequence a practitioner might follow when designing a prompt, reinforcing a practical workflow rather than an abstract checklist.

Beyond these practical benefits, layering also contributes to the broader goal of prompting as a communicative design process. Rather than treating AI as a passive respondent to open-ended queries, layering encourages users to adopt an active stance as authors and architects of interaction. In this sense, prompt layering is not merely a technical function; it is a critical skill that enables precision, transparency, and responsibility in human–AI collaboration.

5.3.1 Diagnosing Weak Prompts and Rewriting Them

One of the central challenges in prompt engineering is recognizing when a prompt underperforms and understanding why. Weak prompts often yield vague, inconsistent, or misaligned outputs not because the model lacks capability, but because the instructions are ambiguous, incomplete, or overly general. Diagnosing and rewriting such prompts requires a shift in perspective: from attributing errors to the system's limits to refining the human input.

A weak prompt usually omits one or more dimensions of control. For example, a request such as "Summarize this" does not guide the desired length, the target audience, the tone of expression, or the intended purpose. The model is therefore forced to make assumptions, often defaulting to generic output. By contrast, a revised prompt such as "Summarize this text in three bullet points for a policy briefing, using formal tone and no more than 100 words" establishes clear boundaries and audience targeting. The difference lies not in verbosity but in informational density and contextual clarity.

Another recurring weakness involves open-ended requests for improvement. A vague instruction like "Improve this paragraph" is unlikely to produce valuable results unless "improvement" is defined. Improvement might mean correcting grammar, clarifying style, simplifying jargon, or adjusting tone; each requires different actions. A more effective alternative would be: "Revise this paragraph to improve clarity and remove technical jargon, while preserving the original meaning." This refined version narrows the task, specifies explicit constraints, and reduces interpretive uncertainty. In doing so, it transforms an open request into a structured instruction that the system can follow more reliably.

Diagnosing weak prompts involves analyzing their clarity across several dimensions. First is the definition of the task, which specifies exactly what the user is asking the model to do. Audience specification determines the vocabulary, tone, and level of assumed knowledge for the content. Users should also define structural expectations, including sentence limits, formatting conventions, and output categories, to ensure consistency and clarity of content. They must clearly delimit the scope to guide the model on which aspects to include or exclude. Ultimately, tone and voice significantly impact whether the output aligns with the intended communicative context.

To support professional practice, a concise verification rubric can be applied when assessing responses. Four checks are handy:

- Format adherence (Does the output follow the required structure?)
- Factual anchoring (Is it consistent with reliable sources?)
- Constraint recall (Are instructions such as tone, word count, or style respected?)
- Coverage vs. brevity (Is the response both complete and succinct?)

This rubric provides a transparent method for evaluating outputs and guides decisions about revision.

A rewriting strategy can then follow a structured formula: instruct the model to perform a defined task, in a specified format, for a given audience, using an appropriate style and tone, under explicit constraints. This approach increases the likelihood of a suitable response while also enabling iteration and improvement. If the model produces unsatisfactory results, users can reassess and refine the prompt using the same parameters.

Rewriting prompts is both an art and a method. In many cases, the failure of a prompt reveals more about the user's assumptions than about the model itself; the act of rewriting surfaces those assumptions, enabling a more deliberate engagement with the AI. Ultimately, diagnosing and refining prompts turns prompting from trial and error into a disciplined practice of communication design.

Later examples in the book progressively show how a weak prompt can be iteratively strengthened by introducing CASTROFF dimensions step by step, allowing readers to see the cumulative effect of each design choice.

5.3.2 Reusability and Prompt Templates

As prompt engineering matures into a professional practice, the need for scalable, repeatable, and auditable methods becomes increasingly evident. Prompt templates address this need by providing structured, reusable formats that can be adapted to multiple contexts while maintaining consistency, clarity, and alignment with communicative goals. Rather than crafting each prompt from scratch, templates provide a modular approach that strikes a balance between flexibility and control.

A predefined structure, known as a prompt template, allows users to insert variable content. It typically contains placeholders for key elements such as the task, audience, tone, output length, and subject matter. For example, a generic template might read: "Summarize {document} for {audience} in {X} words, using {tone} language." Users can adapt this structure for various domains, such as law, education, healthcare, or business, by inserting the relevant parameters. The use of templates allows for rapid deployment of consistent prompts across projects, users, or teams, which is particularly valuable in organizational settings where output must meet institutional standards.

Beyond efficiency, prompt templates also serve a vital ethical and communicative function. Templates integrate tone, format, and audience considerations directly into their structure, encouraging users to exercise greater linguistic responsibility. This approach proves especially important when handling sensitive data, engaging with vulnerable groups, or navigating cross-cultural communication. A well-designed template ensures that responses are not only technically accurate but also socially appropriate.

Canonical schemas illustrate how templates can support professional practice. A JSON report schema defines a strict output with fields such as {title}, {summary}, {data_points}, and {conclusion}, which is valuable for technical documentation or analytics pipelines. A policy brief schema structures content into {issue}, {evidence}, {options}, and {recommendation}, guiding the model to produce outputs that meet policymaking conventions. In healthcare, a patient-safe advice note schema might include {symptom}, {advice}, {safety warning}, and {follow-up}, ensuring clarity while embedding safeguards against harmful ambiguity. These schemas demonstrate how prompt templates can encode professional standards into repeatable linguistic structures.

For instance, in an academic setting, a template may guide the model to "Explain {concept} to {audience level}, using formal tone and including one illustrative example." In a business context, a template might prompt the model to "Compare {product A} and {product B} using three key criteria, then recommend one in a professional tone, limited to 200 words." These examples demonstrate how templates facilitate both functional consistency and domain-specific nuance. They also reduce the likelihood of bias, omission, or stylistic drift by making expectations explicit at the design stage.

Templates further support collaboration by making prompts legible and shareable across users. In professional environments where teams collaborate to generate, audit, or review content, templates serve as a common language, offering transparency about the intent and structure of each prompt. They also strengthen traceability, enabling users to reproduce or refine outputs as needed.

While templates introduce structure, they do not eliminate creativity or adaptability. Instead, they provide a disciplined framework within which variation can occur. A prompt engineer can still iterate, experiment, and refine within the bounds of the template, but with reduced risk of miscommunication or inconsistency. In this sense, templates are not constraints on human agency, but extensions of it, tools that professionalize prompting by aligning design intent with linguistic execution.

As prompting becomes more central to applied AI systems, the use of templates signals a shift from improvisation to intentionality. It represents a move toward shared standards, ethical responsibility, and operational efficiency. In doing so, templates enable AI to function not just as a responsive tool but as a reliable partner in complex, high-stakes human workflows.

5.3.3 Precision Prompting in Practice: Real-World Patterns

The principles of precision prompting, scaffolding, layering, and template design gain their most whole meaning when applied to real-world tasks. Professional domains reveal recurring patterns that show practitioners how to operationalize precision for specific goals. These patterns do not merely reflect stylistic preferences; they represent functional approaches to instructing AI in ways that are consistent, transparent, and aligned with the user's communicative intent.

One typical pattern involves summarizing research for decision-making contexts. For example, a prompt could function as a virtual research assistant, directing the model to analyze a set number of sources on a particular topic, identify key findings, and display them in a comparative format. The prompt may further direct the AI to format the results in a table, focus on particular criteria, and write in a tone suitable for policy advisors. This pattern scaffolds the task into collection, synthesis, formatting, and tone alignment, ensuring that each step produces an interpretable and reusable component.

Another widely used pattern is comparative analysis, which is instrumental in domains such as strategic planning, procurement, and product evaluation. A well-layered prompt may position the model as an analyst and instruct it to compare two options using predefined criteria, articulate the respective strengths and weaknesses, and conclude with a concise recommendation. The prompt might also specify the desired structure, such as an executive summary followed by a point-by-point breakdown, to guide both the content and its presentation. This approach ensures not only factual relevance but also rhetorical alignment with organizational expectations.

Formal correspondence generation is also a recurring domain for precision prompting. In this context, the prompt might guide the model to draft a professional email or letter from a designated role to a specific recipient, covering a particular issue while referencing relevant background information and outlining next steps. Here, tone and clarity are fundamental, as the model must strike a balance between factual accuracy and relational appropriateness. The prompt may include length constraints, specify the use of plain language, and direct the AI to follow conventional formats for the given channel.

Another useful pattern involves adapting content for accessibility. This may include rewriting existing material for a different audience, such as translating technical reports into plain English summaries or simplifying policy documents for community consultation. In such cases, the prompt carefully regulates tone, reading level, and conceptual density to maintain meaning while adapting the style appropriately. These prompts often rely on scaffolding to separate comprehension from rewriting, allowing the model first to understand the content and then express it anew.

Canonical schemas further illustrate how precision prompting supports consistency. A JSON report schema requires outputs with fields such as {title}, {summary}, {data_points}, and {conclusion}, ensuring interoperability with technical workflows. A policy brief schema structures content into {issue}, {evidence}, {options}, and {recommendation}, guiding the AI toward outputs suitable for government or organizational decision-making. In healthcare, a patient-safe advice note schema may specify {symptom}, {advice}, {safety warning}, and {follow-up} to ensure clarity while embedding safeguards against harmful ambiguity. These schemas demonstrate how templates can encode professional standards into repeatable linguistic instructions.

Finally, turn-based scaffolding enhances creative tasks by guiding the model through a structured sequence of ideation, selection, and development. For instance, the prompt may begin by asking the AI to propose several creative directions within a defined genre or theme. Once the user or system selects a preferred direction, the model develops the chosen concept into an outline before producing a draft. This pattern supports collaborative creativity while maintaining control over narrative form, audience, and length.

What unites these diverse applications is their reliance on structured, layered, and intentional prompting. Each pattern reflects a specific use case, but a shared commitment to communicative clarity, ethical responsibility, and reproducibility binds them all. In transitioning from theoretical techniques to applied design, precision prompting becomes not only a method for enhancing AI outputs but also a practical language for guiding collaboration between humans and machines in real-world workflows.

A visual summary of CASTROFF appears in later chapters, where the framework is shown in relation to real-world prompt construction and debugging. This chapter therefore focuses on conceptual grounding and leaves detailed visualization to the applied sections that follow.

5.4 Summary

This chapter established precision prompting as a core literacy in professional AI practice. It showed how granular constraints, scaffolding, layering, and templates convert informal requests into structured instructions that improve clarity, control, and safety. Through practical examples, the chapter demonstrated that precision is not verbosity, but the disciplined use of structure, audience awareness, and constraints. These skills prepare the reader for the CASTROFF framework in the following chapter, where precision becomes the foundation for accountable and reproducible prompt design at scale.

5.5 Self-Assessment Quiz

1. Which of the following prompts best demonstrates layered control of tone, length, and focus?

    a) Describe the role of AI in healthcare.

    b) Write a plain English summary of three AI tools used in UK hospitals for diagnostics, in no more than 150 words.

    c) What are some examples of AI in medicine?

    d) Talk about AI and health technologies.

2. What is the primary advantage of turn-based prompting in professional workflows?

    a) It avoids the need to define format or tone.

    b) It generates longer responses by default.

    c) It allows iterative review at each step before progressing.

    d) It reduces the model's need for external sources.

3. Which of the following best reflects a weak prompt requiring revision?

    a) Summarize this article in 100 words using a neutral tone for a policy audience.

    b) Improve this.

    c) Rewrite this sentence to enhance clarity without altering meaning.

    d) Draft an executive note in plain language, based on the text provided.

4. Why are reusable prompt templates considered valuable in applied AI settings?

    a) They eliminate the need for human input.

    b) They ensure consistency, scalability, and ethical transparency.

    c) They reduce the training data required by the model.

    d) They automatically improve model accuracy.

5. Which example demonstrates precision scaffolding for a structured briefing task?

    a) Write a summary of climate change.

    b) List three causes of climate change.

    c) Outline three causes of climate change, summarize each in 50 words, and present the findings in a comparison table.

    d) Do your best to explain climate change clearly.

CHAPTER 6

The CASTROFF Framework: A Professional Standard for Prompt Engineering

6.1 Why We Need a Standard Prompting Framework

6.1.1 CASTROFF One-Pager

The CASTROFF dimensions summarized in Table 6-1 establish the vocabulary that this chapter develops in detail. What follows builds the case for why such a structured framework is needed, especially in professional and regulated environments where ad hoc prompting cannot suffice.

H. Tavakoli, *Prompt Engineering for Everyone*,
https://doi.org/10.1007/979-8-8688-2338-1_6

Table 6-1.

Dimension	Definition	Good Example	Weak Example	Common Pitfall
Constraints	Explicit limits on length, scope, or format	"Write in 150 words, using bullet points."	"Explain briefly."	Vague limits lead to verbosity
Audience	Intended reader or user context	"Summarize for 12-year-olds."	"Summarize this."	Default generalist voice
Structure	Internal organization of instructions	"First outline causes, then effects."	"Explain causes and effects."	Flat instructions → incoherence
Tone	Emotional/ attitudinal quality of language	"Write in a warm, empathetic tone."	"Write this."	Tone defaults to verbose neutrality
Role	Identity or perspective assigned to AI	"Act as a safety officer."	"Give advice."	Role ambiguity confuses voice
Output Format	Delivery form of response	"Respond in a table with three rows."	"List points."	Unusable prose when structured data is needed
Focus	Thematic priority of response	"Emphasize risks, not benefits."	"Discuss the topic."	Drifting into tangents
Function	Communicative purpose or job of the output	"Critique this for legal compliance."	"Write about this."	Misaligned purpose reduces value

As the use of generative AI spreads across sectors, from education and design to medicine and law, the act of crafting prompts is increasingly treated as a skill, if not yet a discipline in its own right. However, unlike traditional programming or formal writing, prompting remains unstructured and inconsistent. Practitioners often rely on intuition, trial and error, or copied templates rather than principled methods. In the absence of shared standards, the quality of outputs varies widely not only across domains but even within the same user session. This inconsistency undermines both the reliability of AI tools and the trust that users place in them.

The case for a standard prompting framework arises from two converging needs. First, there is a growing demand for professionalization in how prompts are designed, documented, audited, and reused. As AI becomes embedded in regulated environments, such as healthcare, public services, and education, prompts cannot remain ad hoc. Their formulation carries ethical, legal, and social consequences. A poorly phrased prompt in a medical assistant tool, for instance, may result in the omission of critical safety information. Professionals must approach prompt design with the same rigor they apply to other technical and communicative instruments in such contexts.

Second, without a common language or structure for describing prompt strategies, practitioners lack the tools to reflect on, evaluate, and improve their use of these strategies. Educators can only effectively teach or assess concepts that they properly formalize. Students and professionals alike often struggle to identify what went wrong in a weak prompt or why a minor rewording led to dramatically better results. In turn, this gap makes it challenging to build shared repositories, compare use cases, or develop training programs grounded in transferable skills.

CASTROFF addresses this gap. It is not a formula but a structured vocabulary for prompt planning, analysis, and redesign. It draws from principles in user experience design, linguistics, technical communication, and applied ethics. Each dimension of the acronym, Constraints,

Audience, Structure, Tone, Role, Output Format, Focus, and Function, represents a core consideration that shapes how a prompt is interpreted by an AI system and experienced by its user.

The framework helps expert engineers, educators, public sector professionals, designers, and non-technical users alike use AI with precision and care. By breaking down the components of a prompt into inspectable elements, CASTROFF enables us to move from prompting as improvisation to prompting as deliberate, responsible craft.

This framework also builds directly on the **prompt anatomy** concepts introduced in Chapter 2. Whereas anatomy outlined the essential building blocks of a single prompt, CASTROFF expands those foundations into a professional standard that allows prompts to be analyzed, taught, and audited with consistency across contexts. In practice, each CASTROFF dimension maps back to those building blocks but adds precision, scope, and evaluative criteria. It is therefore not a replacement but a formalization and extension of the earlier model, enabling readers to connect what they already know with a structure designed for practice, audit, and teaching.

6.2 C for Constraints

In the context of prompt engineering, constraints refer to the limits within which an AI model must operate to yield proper, safe, and context-appropriate outputs. These constraints may be technical, ethical, domain-specific, or user-defined, and they play a central role in shaping the success or failure of a prompt.

Large language models process linguistic input probabilistically, unlike traditional software that executes hard-coded instructions, requiring users to express constraints *through language* rather than programming. When constraints are vague or absent, the model may default to generic, verbose, or even misleading responses. However, when constraints are precisely articulated, especially within the structure of a prompt, they become a powerful mechanism for steering model behavior.

Constraints in prompting can take many forms. Some specify the length of the output, for example, by instructing the model to respond in fewer than 100 words. Some prompts specify *format or style* requirements, such as bullet points, tables, or a formal academic tone. Some prompts introduce *topical boundaries*, asking the model to focus on specific regions, periods, or knowledge domains. Others define *functional requirements*, such as demanding outputs in JSON format, prohibiting rhetorical language, or avoiding speculation altogether.

Even when these constraints seem simple, their omission can have disproportionate effects. A student requesting a "short explanation" might receive an extended essay unless the prompt explicitly defines brevity. A job seeker who fails to mention "CV-style bullets" might receive prose instead of a usable list. The absence of explicit constraints introduces ambiguity, something language models tend to fill with verbosity or digression.

High-stakes domains, such as healthcare, law, and finance, demand constraint-setting as both an ethical requirement and a professional duty. These fields operate within strict rules, and so must their corresponding prompts. Constraining the model's scope is not merely a stylistic choice; it is a safeguard against producing harmful, misleading, or non-compliant content.

Ultimately, an effective prompt is not simply an output request. It is a negotiated contract. Constraints define the model's role, the user's expectations, and the boundaries of relevance. Human conversation relies on implicit understanding, while human–AI communication requires explicit articulation of every detail. Constraints, when clearly defined, are what transform informal prompting into a disciplined practice.

6.3 A for Audience

In traditional communication, effective speakers tailor their message to suit the audience. The same principle applies in prompt engineering. Every prompt should begin with a clear understanding of who the output is for. Without an explicit sense of audience, the AI defaults to a generalist stance, producing content that is technically correct but often tone-deaf, vague, or inappropriate for the intended reader or use case.

The user must explicitly specify whether the prompt targets a five-year-old, a busy executive, or a fellow researcher, as large language models cannot automatically determine this. Prompt engineers must explicitly incorporate cultural expectations, professional norms, and accessibility needs into the prompt's language, as models lack inherent understanding of these parameters.

Defining the audience clarifies tone, vocabulary, level of detail, and structure. A scientific explanation designed for high school students should avoid jargon and metaphors that presume domain knowledge. A prompt designed for policymakers may need to prioritize brevity, clarity, and actionable recommendations. A business analyst might require tabular summaries or data-driven insights, whereas a novice user may benefit from analogies and step-by-step explanations.

Audience specification can also carry ethical weight. For example, when prompting AI to generate mental health advice, indicating that the recipient is a teenager experiencing anxiety is crucial, not just for tone, but to avoid content that may be medically inappropriate or emotionally harmful. Similarly, educational prompts must consider learners' language levels, cultural context, and prior knowledge.

In professional settings, audience awareness is what separates casual experimentation from reliable AI integration. Proper implementation ensures that AI-generated content meets professional standards for clients, students, patients, or decision-makers, whereas oversight often results in outputs that require extensive revision or withdrawal.

In prompt engineering, the audience is not an afterthought. It is a design parameter. The most effective prompts are those that treat the AI not merely as a content generator, but as a proxy communicator. When the audience is clearly defined, the prompt becomes not only more precise but also more human.

6.4 S for Structure

In human communication, structure helps us follow complex ideas. The same applies when prompting large language models: structured prompts yield structured outputs. Without clear structural cues, even a well-scoped prompt can produce meandering, incoherent, or disorganized responses.

Structure in prompting refers to the internal organization of a task. It includes how the instructions are sequenced, what sections the output should contain, and whether the response should follow a known pattern, such as a numbered list, a table, a three-act argument, or a problem-solution format. Structure forms the foundation for the model's reasoning process and content selection, though it differs from the output format (covered in Section 6.7).

For instance, asking an AI to "analyze this data" may yield a summary, a critique, or a prediction, depending on how the model interprets "analyze." However, if the prompt specifies: *"First summarize the trends in plain language, then provide a bullet-point list of risks, followed by a three-sentence recommendation,"* the model has a structure to follow. This scaffolding does not limit the AI; it guides it.

Structured prompts are especially powerful for reasoning tasks. Chain-of-Thought prompting, for example, works because it imposes a stepwise structure that mirrors human logic. Similarly, Tree-of-Thought and Self-Ask prompting strategies (discussed in Chapter 4) rely on internal prompt structures that cue the model to diverge, evaluate, or self-question before concluding.

In real-world practice, structure is what allows prompts to scale. Teams that use AI for report writing, coding, design, or customer service often rely on templates or modular structures to ensure quality and repeatability. Without structure, even strong instructions can result in output that feels arbitrary or inconsistent.

For human users, structure increases interpretability. For AI systems, it increases predictability. In prompt engineering, structure is what transforms vague intent into reproducible performance.

6.5 T for Tone

Tone refers to the manner or attitude expressed in language, as well as the subtle cues that frame how information is received, interpreted, and valued. In human communication, tone often carries more weight than content alone. The same principle applies in human–AI interaction. In prompt engineering, tone determines not just what the model says, but how it says it, shaping the emotional, social, and professional resonance of the output.

Language models are capable of mimicking a wide variety of tones, from friendly and conversational to formal, urgent, or neutral. However, without explicit tonal guidance, the model often defaults to a generalist voice, usually neutral, somewhat verbose, and cautiously optimistic. While this may suffice for basic tasks, it can lead to misalignment in professional, educational, legal, or therapeutic contexts where tone must accurately reflect domain-specific norms or user sensitivities.

Imagine a scenario where the AI creates a medical information leaflet for patients who have just been diagnosed with a chronic condition. A prompt that fails to define tone might produce text that is either overly technical or inappropriately upbeat. By contrast, a prompt that specifies a "reassuring and empathetic tone, using accessible language appropriate for a non-specialist audience" is more likely to yield output that is ethically sound and emotionally attuned. Here, tone becomes a form of care, one that extends beyond correctness to consider human impact.

Tone also plays a critical role in perceived authority and credibility. A legal brief requires a restrained, objective tone, whereas a motivational speech may require persuasive energy and rhetorical emphasis. Mismatched tone can undermine the communicative intent, even when factual content is correct. In public-facing documents, this can result in reputational risk or audience disengagement. In sensitive contexts, such as grief support, gender-based violence, or immigration services, a poorly calibrated tone may even cause harm.

Professionals must deliberately calibrate tone by both specifying their desired tone in the prompt and verifying whether the model's output accurately and consistently maintains it. It is essential to remember that tone is not a static feature, but rather a contextual one. A single user may require different tones across tasks, such as a friendly tone in onboarding materials, a formal tone in compliance documentation, or a neutral tone in data reporting.

From an ethical standpoint, tone influences trust. Appropriate, respectful, and well-judged AI-generated content increases user engagement and action, particularly in multicultural or multilingual contexts where tonal nuances differ substantially. Prompt engineers must therefore account for cultural sensitivities, organizational norms, and interpersonal expectations when specifying tone.

To support this, prompts should include explicit tonal cues where relevant, for example: "Write a two-paragraph summary in a formal and respectful tone for a senior policymaker," or "Explain this to a teenager using a supportive and encouraging tone." These instructions serve both a practical and normative function: they help the model produce suitable language while reinforcing the human values embedded in the task.

In sum, tone is not cosmetic. It is a core communicative variable that mediates meaning, shapes perception, and governs relational trust. In precision prompting, tone must be intentionally specified, critically evaluated, and ethically designed. When used effectively, it not only enhances the clarity of the message but also the dignity of its reception.

6.6 R for Role

In prompt engineering, the concept of role refers to the communicative identity assigned to the AI system within a task. It determines the perspective, voice, and assumed expertise from which the model responds. While language models do not possess identity in any literal or psychological sense, they are highly responsive to role-based cues embedded in prompts. These cues significantly shape the tone, structure, and epistemic framing of the output.

Role-based prompting does not merely set a mood or stylistic preference; it governs the model's rhetorical positioning. In professional contexts, this determines whether the AI acts as an explainer, advisor, editor, analyst, or collaborator. For example, asking a model to "act as a science teacher explaining climate change to a 14-year-old" produces markedly different output from a prompt that requests the same explanation "as a policy advisor writing for government officials." The knowledge depth, terminology, emphasis, and presentation style shift accordingly.

Assigning a role helps the prompt engineer replicate domain-specific expertise or a communicative perspective, proving particularly valuable in fields that require clarity, consistency, or authority, such as legal, educational, or healthcare contexts. For instance, a model cast in the role of a safety officer reviewing a workplace procedure will tend to foreground risks and compliance. In contrast, a marketing strategist in the same context may emphasize presentation and engagement. In each case, role cues guide the model to retrieve and prioritize relevant patterns from its training data, narrowing interpretive ambiguity.

Well-defined roles also enhance ethical and functional precision. In high-stakes or emotionally sensitive applications, role specification ensures that the model's language aligns with professional responsibilities. A prompt asking the model to "act as a supportive peer offering encouragement to someone with exam anxiety" introduces a fundamentally different frame from a clinical or diagnostic approach. Here, the choice of role acts as a boundary on the kind of content that is appropriate, responsible, and contextually aware.

It is essential to distinguish between role and tone. While tone relates to how something is said, role determines the underlying purpose and perspective. A single role, such as a historian, can operate across multiple tones (e.g., formal, conversational, critical). The same neutral tone can effectively serve multiple roles, whether as a journalist or mediator. Effective prompting depends on the ability to calibrate both dimensions in tandem.

From a pedagogical perspective, role-based prompting also supports learning and exploration. Casting the model as a simulated tutor, reviewer, coach, or peer enables users to explore unfamiliar material with guidance and support. This approach particularly benefits non-experts seeking customized explanations or professionals simulating alternative viewpoints and stakeholder positions. In such cases, the role becomes not just a control parameter but a tool for perspective-taking and dialogue.

However, there are limits to role fidelity. A language model cannot truly reason like a lawyer or provide emotional support like a therapist. Role simulation relies on linguistic approximation, not lived experience or institutional training. As such, role-based prompts must be crafted with care and evaluated critically, particularly when used in domains that require certification, duty of care, or cultural sensitivity.

In summary, the role dimension of CASTROFF offers a mechanism for narrowing scope, aligning expertise, and shaping the perspective of AI-generated content. When well-specified, roles reduce ambiguity, enhance relevance, and support ethical clarity. In professional-grade prompting, assigning a role is not an embellishment; it is a design strategy that brings coherence, trustworthiness, and contextual intelligence to human–AI interaction.

6.7 O for Output format

The output format determines the external presentation structure of the AI's response. While structure (Section 6.4) determines the logical arrangement of ideas, output format controls their visual or textual presentation, including options like lists, tables, paragraphs, code snippets, email drafts, slide outlines, or other artifact types. In prompt engineering, explicitly specifying output format enables not just control over clarity and readability but also alignment with the expectations of a given task, domain, or workflow.

When the output format is omitted or left implicit, the language model defaults to generic prose, typically in paragraph form, often overly verbose. While this may be acceptable for informal interaction, it is often inadequate in professional settings where content must be scannable, structured for action, or inserted directly into another system. For instance, in technical documentation, an unordered list of functions may be more usable than a narrative description; in legal practice, bullet points summarizing clauses can support rapid review; in education, a tabulated comparison may better highlight key distinctions.

Specifying output format also increases interoperability. Applied workflows often incorporate AI-generated outputs that demand exact formatting specifications. A business analyst may need model outputs in CSV format to support data imports. A programmer may request JSON for use in APIs. A communications officer may want output formatted as markdown for internal documentation. Without clearly articulated format constraints, AI systems may generate content that is stylistically correct but structurally unusable.

Moreover, format specification often enhances accessibility. Users with cognitive or visual impairments may benefit from content rendered in bullet points or numbered steps. Prompts that request clear formatting, such as headings, emphasis, or line breaks, can reduce cognitive load and make content more inclusive. In such cases, format is not only functional; it is an equity concern.

In prompt engineering, it is also helpful to distinguish between format as delivery and format as instruction. Output format determines how the response appears to the reader or system, while role assignment guides the model's behavior within a specific communicative context. For example, "as a bulleted list" is a delivery instruction. At the same time, "provide three advantages and three limitations, each on a new line" guides both the structure and formatting of content. Effective prompts often do both: they define the format for readability and specify structural expectations for completeness.

However, overly rigid format constraints can occasionally limit the model's ability to respond fluidly, especially in creative or exploratory tasks. Effective prompting strikes a balance between providing precise instructions and allowing for adaptable language use. In high-stakes settings, erring on the side of specificity is often prudent. In low-stakes or generative contexts, looser formatting may support better creative exploration. The decision should reflect the communicative purpose of the task and the cognitive needs of the user.

In summary, output format is not merely a matter of presentation; it is a critical dimension of prompt engineering that influences clarity, usability, accessibility, and interoperability. By specifying output format explicitly, whether as a visual guide, a data container, or a rhetorical shape, prompt engineers increase the reliability, efficiency, and professionalism of AI-generated responses. Within the CASTROFF framework, format functions not as an afterthought but as a delivery scaffold, connecting intention to implementation in human–AI communication.

6.7.1 F for Focus

Focus refers to the conceptual boundaries that define what the AI should prioritize in its response. Constraints (Section 6.2) define operational limits such as word count, tone, or style, while focus determines the cognitive perspective that frames the task. It instructs the model to emphasize particular aspects of a topic while ignoring or deprioritizing others. In doing so, focus shapes not just the content of the output but its interpretive center.

Language models are capable of retrieving and combining a wide range of information. Without clear focus cues, however, the output may become diffuse, covering tangential topics, repeating ideas, or presenting content that is technically accurate but strategically misaligned. A prompt that asks for "an overview of renewable energy" may yield a disorganized mixture of technologies, global statistics, and policy debates. A more focused prompt, such as "summarize the economic barriers to solar energy adoption in sub-Saharan Africa for a policy audience," creates a sharper frame. It defines scope, audience, and subject emphasis, all of which shape the informational intent of the response.

Thematic, procedural, or analytical dimensions establish the focus of a prompt. Thematic focus instructs the model to concentrate on specific content areas, such as causes without solutions, risks without benefits, or implications without context. A procedural focus might limit attention to specific stages of a workflow, such as data collection but not analysis, or diagnosis without treatment. Analytical focus prioritizes forms of reasoning, such as causal explanation, comparison, critique, or interpretation. Each of these approaches can be embedded directly in the language of a well-engineered prompt.

The importance of focus becomes especially evident in professional and institutional settings, where users often require the model to serve a specific communicative purpose. A research assistant may be asked to present only contradictory findings on a specified topic. A legal consultant may request content that highlights regulatory risks while excluding descriptive commentary. A civic engagement officer might need a version of a document that foregrounds community concerns while omitting technical language. In each case, the goal is not to restrict knowledge but to direct attention toward relevance, clarity, and communicative utility. As discussed in Chapter 2's *prompt anatomy*, focus works alongside constraints to establish not only the limits of a task but also its intellectual center.

From an ethical standpoint, focus also plays a critical role in transparency. Unfocused prompts may yield outputs that feel comprehensive but obscure bias, omit key viewpoints, or misrepresent uncertainty. A prompt that fails to guide the model toward the most relevant, timely, or inclusive information can inadvertently reproduce dominant perspectives while marginalizing others. Prompt engineers must therefore treat focus as both a strategic and normative concern. What the model includes and what it leaves out reflects the framing decisions made by the human designer.

It is essential to distinguish between focus and function. Function defines why the output is being produced, whether to inform, persuade, analyze, or advise. Focus specifies what content receives attention within that purpose. For example, the function of a policy brief may be to inform legislators, but its focus could be limited to economic risks rather than providing detailed technological information. Separating the two ensures that prompts are both relevant to the task and precise in what they emphasize. Without this distinction, outputs risk being broad but shallow, or targeted but misaligned with the intended use.

Focus also differs from completeness. A focused response maintains depth within its specified parameters. One of the virtues of precision prompting is that it enables depth without breadth, allowing users to explore issues with greater rigor by isolating them from surrounding noise. For this reason, prompts that explicitly name their focal concern tend to produce more useful, interpretable, and context-aware outputs.

In summary, focus determines the thematic or analytical core of a prompt. It sharpens the AI's attention, aligns output with user intent, and promotes ethical awareness by making framing decisions transparent. Within the CASTROFF framework, focus is not a narrowing of possibilities but a clarification of purpose. It enables prompt engineers to move beyond generalized content toward meaningful, directed communication.

6.7.2 F for Function

Function refers to the intended use or communicative purpose of the AI-generated output. Tone, role, and format influence the delivery of content, whereas function establishes the underlying reason for creating the content. In prompt engineering, this dimension grounds the task in a real-world goal, whether it is to inform, persuade, instruct, evaluate, summarize, critique, or advise. A clearly defined function prevents technically accurate but contextually inappropriate outputs by aligning the model's behavior with the wider communicative purpose.

A prompt that asks for "a summary of recent research on AI in education" leaves the function open to interpretation. Is the summary intended for policymakers, students, educators, or funders? Is the goal to inform a decision, to support teaching, or to shape public understanding? Each of these purposes would require a different form of emphasis, vocabulary, and structure. Without functional framing, outputs risk being vague, misplaced, or even counterproductive.

Embedding a function directly into the prompt, through verbs that define purpose, helps the model select and prioritize information more effectively. Prompts that begin with "Summarize to inform a public consultation," "Critique this proposal from a legal standpoint," or "Generate questions to test comprehension" immediately set a direction. They define the output's purpose rather than merely its appearance, creating more actionable prompts and valuable results.

In applied settings, function also interacts with regulatory and ethical demands. For example, a prompt intended to generate clinical information must serve a different function than one designed to support emotional reassurance. A diagnostic tool must prioritize precision and caution, while a helpdesk assistant must balance efficiency with clarity. Misidentifying function, or failing to specify it, can result in inappropriate tone, content omissions, or unintended consequences.

Function enhances auditability and facilitates human review by enabling evaluators to assess whether the content fulfilled its designated purpose. A well-specified function makes it easier to judge the relevance, accuracy, and completeness of the results. Clear functional intent in prompt design also promotes reuse across similar use cases, organizations, or tools.

It is also important to note that multiple functions may coexist in a single prompt. A prompt may simultaneously aim to inform and persuade or to instruct while also building confidence. Prompt designers must determine which function holds priority and convey this hierarchy in such cases. Making this decision explicit in the prompt helps the model navigate complex expectations with greater coherence and consistency.

Function differs from focus. Focus determines what content the AI should emphasize within its response (e.g., economic barriers, community concerns, or comparative risks). Function defines what the output is ultimately for, its role in human communication (e.g., informing, persuading, advising). A single focus may serve different functions, and a single function may employ different focuses. Maintaining this distinction ensures prompts retain precision in emphasis and purpose in outcome.

Finally, function serves as the bridge between design intent and human need. It reminds us that AI-generated outputs are not ends in themselves but tools in a broader communicative chain. The model does not simply produce language; it participates in a task with goals, implications, and consequences. Within the CASTROFF framework, function anchors the prompt in purpose. This approach elevates prompting from a technical exercise to purposeful design, aligning language, structure, and ethics with human goals.

6.7.3 Clarifying Note: Focus vs. Function

- **Focus** sets the content boundaries: what to emphasize, highlight, or leave out. It controls scope and depth.
- **Function** defines the job the output must perform: explain, persuade, summarize, instruct, reassure, etc.
- Together, focus ensures the right material is included, while function ensures it is delivered in the right way for the task.

Example

- **Focus:** "Summarize causes of climate change." → Output emphasizes greenhouse gases, deforestation, and energy use.
- **Function:** "Write a public awareness leaflet on climate change." → Output is short, persuasive, and calls to action.

Same subject, but different outcomes. Focus narrows what is covered; function directs how it serves the reader.

6.7.4 Rapid Triage Flowchart: Which CASTROFF Knob to Turn?

When a prompt goes wrong, start with the symptom. Adjust the CASTROFF dimension that best matches the problem:

- **Output is too long → tighten constraints** (e.g., word count, bullet points).
- **Output is too short/thin → expand focus** (add scope, details, or examples).

- **Response drifts off-topic → sharpen function** (state the job clearly).
- **Tone feels wrong → adjust tone** (supportive, neutral, formal, plain).
- **Style is inaccessible/confusing → fix structure** (steps, headings, lists).
- **Model ignores role/audience → reset role or audience.**
- **Answer is verbose/technical → simplify output format** (plain English, CEFR B1/B2).
- **Information feels unbalanced → refine focus** (emphasize what matters).

Think of CASTROFF as a control panel: minor adjustments often repair prompts faster than rewriting from scratch.

6.7.5 Recap: The CASTROFF Framework

This chapter introduced the CASTROFF framework as a professional standard for analyzing, designing, and evaluating prompts in human–AI communication. CASTROFF provides a structured vocabulary for prompt engineering, offering eight interrelated dimensions: constraints, audience, structure, tone, role, output format, focus, and function. Each dimension enables greater clarity, control, and ethical alignment in the construction of AI instructions.

Together, these elements transform prompting from a trial-and-error activity into a deliberate, inspectable process. Constraints define operational boundaries such as length, scope, or format. Audience awareness tailors the output to meet the specific needs, background, and expectations of a particular group. Structure provides internal organization that guides the model's reasoning process, while tone shapes the emotional and interpersonal quality of the language. The role assigns the AI a communicative perspective or identity, such as tutor, analyst, or peer, which informs the framing and authority of the output. Output format controls the delivery form, whether prose, list, table, or code, and ensures compatibility with workflow or accessibility requirements. Focus sharpens the model's attention on relevant aspects of the task, while function clarifies the real-world purpose behind the generation.

While each CASTROFF dimension is independently meaningful, their combined use is what enables prompt engineering to support professional-grade outputs. In practice, they act not as isolated parameters but as overlapping design principles. A change in tone may influence audience fit; a shift in structure may support or undermine the intended function. Understanding these relationships enables prompt engineers to diagnose weak prompts, repair misalignments, and design with greater intentionality.

As AI becomes embedded in systems of communication, education, governance, and care, the need for structured prompting frameworks becomes critical. CASTROFF meets this need by offering both conceptual clarity and practical applicability. It is not a formula to follow rigidly, but rather a language for thinking and discussing the quality, safety, and purpose of prompts.

The following chapter demonstrates CASTROFF's practical application through diagnostic analysis, prompt refinement, team collaboration, and real-world case studies. By mastering the CASTROFF framework, practitioners move one step closer to building responsible, transparent, and effective AI–human interactions.

6.8 Self-Assessment Quiz

1. What is the primary function of constraints in a well-formed prompt?

   a) To reduce the overall length of the model's response

   b) To prevent the model from hallucinating completely

   c) To limit scope and clarify expectations

   d) To enhance the emotional tone of the output

2. Why is it important to specify an audience in prompt design?

   a) It instructs the model to use internet sources.

   b) It defines the role the model should adopt.

   c) It calibrates the tone, relevance, and complexity of the response.

   d) It compresses the output into a smaller file size.

3. How does the structure differ from the output format in the CASTROFF framework?

   a) Structure affects tone, while format controls word count.

   b) Structure refers to internal layout; format determines file type or medium.

   c) Structure changes the AI's personality; format determines sentence length.

   d) Structure and format are functionally equivalent.

4. When a user asks the AI to act as a policy analyst, which CASTROFF element is being applied?

   a) Role

   b) Focus

   c) Function

   d) Audience

5. What is the purpose of adding focus to a prompt?

   a) To ensure perfect grammatical output

   b) To draw the model's attention to specific priorities or interpretive angles

   c) To prevent bias in all circumstances

   d) To generate shorter, more concise responses

CHAPTER 7

Applying CASTROFF in Practice: Diagnosing, Repairing, and Redesigning Prompts

7.1 Diagnosing Weak Prompts Using CASTROFF

Diagnosing a weak prompt is the first step toward professional prompt engineering. It marks the shift from passive interaction with AI systems to active design responsibility. A poorly performing prompt is not necessarily a reflection of model limitations; more often, it reveals a mismatch between the user's intent and the clarity or structure of the instruction. The CASTROFF framework enables users to identify and articulate the root causes of this mismatch, offering a diagnostic lens that replaces vague intuition with analytical precision.

H. Tavakoli, *Prompt Engineering for Everyone*,
https://doi.org/10.1007/979-8-8688-2338-1_7

Many users experience prompt failure as a kind of communicative friction: outputs are too long, too vague, off-topic, overly technical, or misaligned with their professional needs. Users frequently dismiss these breakdowns as isolated frustrations, addressing them through repeated experimentation. CASTROFF reframes this experience as a structured opportunity for design reflection. Each of its eight dimensions, constraints, audience, structure, tone, role, output format, focus, and function, offers a different point of entry for understanding what has gone wrong and why (see Chapter 6 for a complete introduction to these categories).

For example, if an AI-generated response is verbose and unfocused, the underlying issue may relate to weak constraints and a lack of specified focus. If the tone feels inappropriate or overly casual, the problem may lie in the absence of tone guidance or unclear role specification. If the output is factually correct but pragmatically unhelpful, the function that the user intends the model to do was never made explicit. CASTROFF does not treat these as abstract qualities but as repairable design parameters.

Diagnosing weak prompts involves asking a series of interpretive questions, many of which map directly to CASTROFF's eight dimensions. What are the implicit or missing constraints? Is the intended audience defined or assumed? Has the structure of reasoning been suggested or left open? What tone is the AI adopting, and does it match the task or audience? Is the model assuming a helpful role or merely echoing a neutral narrator? Has the output format been specified in a way that supports usability? Has the focus been narrowed to the relevant aspects of the problem? Finally, what is the communicative function of the output: informing, persuading, advising, or something else?

By engaging with these questions, prompt engineers move from surface-level editing to structural diagnosis. Rather than asking the AI to "do better" or "try again," they begin to articulate the specific ways a prompt has failed to scaffold clarity, purpose, or ethical alignment. In professional environments, whether legal, educational, policy-based, or healthcare-related, this kind of diagnostic rigor becomes essential. Weak prompts not only reduce efficiency but also introduce risk, misrepresentation, and a loss of trust.

Importantly, CASTROFF-based diagnosis also supports pedagogy and collaboration. It provides a shared language for peer review, group audits, and institutional prompting standards. In team-based settings, it allows multiple stakeholders to critique and refine prompts without needing programming expertise. It elevates the act of revision from stylistic tweaking to strategic communication design.

To aid this process, later sections of this chapter introduce structured exercises that ask readers to annotate and revise flawed prompts using the CASTROFF lens. These activities reinforce the idea that prompting is not simply a matter of asking questions but of designing interactions with specific goals, boundaries, and users in mind. Just as a software engineer debugs code, a prompt engineer diagnoses language, locating the friction, tracing it to its cause, and refining the interaction for accuracy, clarity, and utility.

7.2 Prompt Repair Workshop: Everyday Scenarios

Following the CASTROFF-based diagnosis of a weak prompt, practitioners should implement focused revisions. This section offers a series of applied scenarios that model the process of prompt repair, not through superficial rewording, but by aligning each revision with one or more CASTROFF dimensions. Each example moves beyond a "before-and-after" format to demonstrate how structural and ethical reasoning inform the design of professional prompts.

Prompt repair is both a linguistic and conceptual exercise. Effective prompt improvement requires evaluating multiple dimensions, including task clarity, tone suitability, format functionality, audience targeting, and alignment with purpose, rather than merely linguistic edits. This diagnostic process requires examining both the AI's shortcomings and how the original prompt inadequately directed the model's response.

7.2.1 Scenario 1: Ambiguous Purpose and Structure

Original Prompt

"Can you help me write something about climate change for a school?"

Revised Prompt

"Write a 300-word summary on the causes of climate change for UK secondary school students, using plain language, a neutral tone, and one real-world example."

7.2.2 Repair Analysis

The original prompt lacks constraints (length), audience specification (age or reading level), role (informational vs. persuasive), structure, and function. It invites a broad, unpredictable response. The revision imposes explicit length and audience constraints, shifts the tone to neutral, and clarifies that the purpose is to inform, not argue. This approach demonstrates how the CASTROFF dimensions (constraints, audience, focus, tone, and function) can undergo simultaneous repair through meticulous re-specification.

7.2.3 Scenario 2: Tone Mismatch in Sensitive Contexts

Original Prompt

"Write a message to someone who just lost their job, telling them to stay positive."

Revised Prompt

"Compose a short, empathetic message offering support to a colleague who has recently been made redundant. Avoid clichés and maintain a tone of respect and solidarity."

7.2.4 Repair Analysis

The original prompt risks being emotionally tone-deaf, despite good intentions. It directs the model toward an upbeat tone without attending to emotional sensitivity. The revised version names the tone explicitly (empathetic, respectful), provides contextual clarity (colleague, redundancy), and discourages inappropriate language ("avoid clichés"). Here, tone, audience, and role are all rebalanced to prioritize relational ethics over optimism.

7.2.5 Scenario 3: Missing Output Format in a Professional Setting

Original Prompt

"Compare renewable energy sources."

Revised Prompt

"Create a two-column table comparing solar, wind, and hydroelectric energy based on cost, availability, and environmental impact, using bullet points and concise phrases suitable for a policy briefing."

7.2.6 Repair Analysis

The revised prompt establishes explicit constraints by introducing a table format, defining specific comparison criteria (cost, availability, impact), and specifying the professional context (policy briefing). This transformation repositions the model from a generalist to a focused summarizer, with explicit output format, structure, and role parameters that yield more practical and accessible results.

7.3 Comparing Before and After AI Outputs

While prompt repair focuses on constructing better instructions, evaluating outputs remains an essential complementary skill. Practical prompt engineering is not simply about generating revised prompts; it is about critically assessing the results that prompts produce. This section illustrates how CASTROFF can be used to compare before-and-after outputs, providing a structured vocabulary for determining whether a revision improves clarity, relevance, and fitness for purpose.

An effective revised prompt generates responses that improve not only in accuracy and style but also in alignment with human intention, ethical considerations, and communicative purpose. To assess this, prompt engineers must move beyond surface preferences and develop a principled way to compare outputs. The CASTROFF dimensions offer such a framework, enabling evaluators to ask: Did the revised prompt produce output with more explicit constraints? Was the intended audience better reflected in tone or terminology? Did the model adopt the desired role or format? Was the focus tighter or the function more transparent?

Consider the following comparison. A user initially writes the following:

Initial Prompt

"Tell me about AI in education."

Initial Output (Excerpt)

"AI is changing education in many ways. It can help teachers grade more efficiently, support personalized learning, and reduce administrative burdens. However, there are concerns about bias, privacy, and the loss of human interaction."

This response is factually accurate and stylistically neutral; however, its scope is unfocused, and it lacks context, structure, and audience adaptation. The revised prompt reads as follows:

Revised Prompt

"Prepare a 200-word briefing for UK education policymakers analyzing AI's potential benefits and implementation challenges in secondary schools, using accessible language with a formal register."

Revised Output (Excerpt)

"Artificial intelligence (AI) offers significant benefits to UK secondary schools, including automated grading, learning analytics, and personalized instruction. These tools can increase efficiency and tailor support to student needs. Nevertheless, significant challenges persist, particularly regarding data protection concerns, potential algorithmic bias, and the risks of excessive reliance on automated systems. Policy responses must balance innovation with safeguards to protect equity and trust."

This second output is more targeted, more informative for the intended audience, and more structurally consistent with policy communication. The differences are not cosmetic. They reflect how CASTROFF dimensions, specifically audience, tone, focus, structure, and function, were made explicit in the prompt and thus internalized by the model during generation.

Such comparisons reveal that prompt engineering is not merely iterative; it is evidentiary. Revising a prompt is an act of design, but evaluating its output is an act of interpretation and justification. In professional practice, this may involve not just individual judgment but collaborative review. Teams should choose the most suitable output for publication, substantiate their selection using CASTROFF principles, or select between functionally different versions based on specific audience needs and task requirements.

Moreover, comparative evaluation helps surface unintended consequences. A revised prompt that improves focus may inadvertently narrow inclusivity. A format change may enhance usability while flattening nuance. CASTROFF helps prompt engineers to articulate these trade-offs and make them visible in practice.

In short, comparing before-and-after AI outputs is not about proving one version is objectively better than another. It is about learning *how and why* specific dimensions of a prompt affect the quality and appropriateness of machine-generated language. It is also about developing critical literacy: the ability to explain, rather than merely prefer, a design choice.

In the following section, we extend this evaluative process into collaborative contexts, examining how teams can review, critique, and improve prompts together through structured audits and peer feedback.

7.4 Team Prompt Audits and Peer Feedback

Prompt engineering is rarely a solitary activity. The integration of AI tools into professional sectors, such as education, policy, healthcare, and journalism, transforms prompt design, refinement, and evaluation into collaborative endeavors. Whether in design teams, editorial boards, curriculum groups, or technical task forces, prompts are shaped through deliberation, not intuition. This section introduces the practice of team prompt audits and peer feedback as essential steps in building organizational prompting standards and shared communicative accountability.

A team prompt audit involves the structured evaluation of prompts, either live or retrospective, using the CASTROFF framework as a diagnostic and discussion tool. The process may begin with a single team member presenting a prompt that underperformed or produced problematic results. The group then collaboratively reviews the prompt by considering how each CASTROFF dimension was applied or omitted. What constraints were missing or unclear? Was the intended audience specified or assumed? Did the tone fit the context? Were role, structure, or format well-articulated? What was the function of the output, and did the prompt support it?

This collective review does more than refine individual prompts. It trains teams to think critically and communicatively, using a shared vocabulary to unpack both technical and ethical implications. For instance, a prompt that asks the model to “generate public-facing messages on vaccine safety” may prompt discussion about tone calibration (reassuring vs. assertive), audience clarity (parents vs. clinicians), and function (informing vs. persuading). CASTROFF provides the conceptual scaffolding to support these discussions, enabling them to move beyond opinion into principled design critique.

Educational and institutional settings can embed peer feedback through rubrics or audit checklists, requiring participants to evaluate colleagues’ prompts using CASTROFF criteria to identify both practical elements and opportunities for improvement. These exercises are not only pedagogically practical; they also model the kind of professional review processes increasingly needed in AI-integrated workplaces. Additionally, they demystify prompt evaluation, showing that it is not the domain of technologists alone, but also of educators, analysts, communicators, and designers.

Importantly, team audits foster epistemic diversity. Different users bring different perspectives to a prompt: what sounds transparent or ethical to one person may feel exclusionary or ineffective to another. Collaborative evaluation reveals these variations, allowing teams to redesign prompts with enhanced inclusivity and cultural sensitivity. In this sense, prompt audits are not just about quality control; they are also about institutional learning and accountability.

For teams working with public-facing or high-stakes AI outputs, peer review may also act as a safety mechanism. Misaligned tone, inappropriate content, or structural ambiguity can be identified before deployment, thereby reducing reputational, operational, and ethical risks. In sectors such as government, healthcare, or journalism, this form of pre-deployment scrutiny is rapidly becoming a necessity.

Finally, prompt audits provide a means to develop prompting literacy across various roles and departments. By making the CASTROFF framework explicit and collaborative, organizations can move toward shared prompting standards, grounded not only in efficiency but also in clarity, equity, and professionalism.

In the final section of this chapter, we turn from diagnosis and peer review to the complete redesign of complex prompts. Through detailed case studies, we explore how to move from weak or ambiguous instructions to robust, context-sensitive, and ethically sound prompt architecture.

7.5 Full Prompt Redesign Case Studies

This final section demonstrates the complete redesign of prompts in high-context, real-world scenarios. Each case begins with an underperforming prompt that reflects a standard failure mode, such as ambiguity, inappropriate tone, structural incoherence, or ethical misalignment. CASTROFF guides the redesign process, transforming each prompt into precise, intentional instructions aligned with communicative objectives and situational requirements.

These case studies move beyond isolated fixes to illustrate how multiple CASTROFF dimensions interact in practice. They also reveal the depth of reasoning required to create prompts that are not only functional but also sensitive to audience needs, ethical responsibilities, and domain-specific language practices.

7.5.1 Case 1: Public Service Form Helper

Original Prompt

"Help me with this housing form."

Issues Identified

The request is vague, with no information about the form type, target audience, or task scope. The model could generate irrelevant or incomplete advice, risking confusion in a sensitive public service context.

Redesigned Prompt

"Act as a housing adviser. Rewrite the instructions for completing Section 3 of the UK Housing Benefit form in plain English, limited to 200 words. Use numbered steps, avoid legal jargon, and include one short example."

7.5.2 CASTROFF Dimensions Applied

- **Role**: Housing adviser
- **Audience**: Applicants with a limited technical background
- **Structure and format**: Numbered steps, example
- **Constraints**: 200-word limit, plain English
- **Function**: Clarification and guidance

Resulting Output

The revised prompt generated clear, step-by-step instructions, using accessible vocabulary and a short illustration.

7.5.3 Measurable Deltas

- **Token reduction**: 41% fewer tokens than the original attempt
- **Format adherence**: 100% (numbered list delivered as requested)
- **Hallucination checks**: Passed (no fabricated details, verified against form)

7.5.4 Case 2: HR Email

Original Prompt

"Write an email to staff about remote work."

Issues Identified

The purpose is underspecified: no details on policy, tone, or required structure. The model could produce either an overly casual or excessively formal message, risking miscommunication.

Redesigned Prompt

"Draft a 150-word HR email to all staff announcing a new hybrid-working policy. Use a professional but supportive tone, include the start date, and summarize expectations in three bullet points. Conclude with contact details for further questions."

7.5.5 CASTROFF Dimensions Applied

- **Role**: HR officer
- **Audience**: Employees across departments
- **Structure and format**: Email with bullet points
- **Constraints**: 150 words, supportive tone
- **Function**: Announcement and clarification

Resulting Output

The redesigned prompt yielded a concise, structured email that balanced professionalism and warmth, with clear action points.

7.5.6 Measurable Deltas

- **Token reduction**: 36% fewer tokens

- **Format adherence**: 100% (bullet points included)
- **Hallucination checks**: Passed (policy details remained generic unless specified)

7.5.7 Case 3: Undergraduate Feedback Note

Original Prompt

"Give feedback on this essay."

Issues Identified

The instruction lacks criteria, a word limit, or guidance on tone. The model may produce unfocused or unconstructive comments, failing to meet academic standards.

Redesigned Prompt

"Provide a 200-word feedback note on this undergraduate history essay. Highlight two strengths and two areas for improvement. Use an encouraging academic tone, and suggest one concrete revision strategy for the next draft."

7.5.8 CASTROFF Dimensions Applied

- **Role**: Lecturer
- **Audience**: Undergraduate students
- **Structure and format**: Short note with balanced points
- **Constraints**: 200 words, constructive academic tone
- **Function**: Formative feedback

Resulting Output

The redesigned prompt produced feedback that was balanced, actionable, and aligned with academic conventions, without generic filler.

7.5.9 Measurable Deltas

- **Token reduction**: 29% fewer tokens
- **Format adherence**: 95% (strengths and weaknesses clearly separated)
- **Hallucination checks**: Passed (all feedback tied directly to essay text)

7.5.10 Reflections on Redesign Practice

These cases illustrate that prompt revision is not only about correcting local errors but also about reimagining communicative intent. Effective redesign uses CASTROFF not as a checklist but as an integrated framework for building purposeful, inclusive, and context-aware instructions. The measurable improvements, token efficiency, format reliability, and factual accuracy demonstrate that structured prompting enhances both technical performance and social responsibility; the strength of a prompt lies in its alignment with human goals and values.

The subsequent chapter expands these principles to collaborative redesign processes and sustainable governance frameworks, employing CASTROFF as the cornerstone for establishing institutional prompting standards.

7.5.11 Integrating CASTROFF into Practice

To transition from individual case redesigns to systematic team practice, teams can implement CASTROFF through a straightforward audit process. The following one-page sheet provides a reusable framework for reviewing prompts, ensuring that each dimension is explicitly considered before deployment.

7.6 CASTROFF Audit Sheet

This tool supports individual or team-based audits of prompts. It can be photocopied, downloaded, or adapted for training. Each dimension includes a tick-box for presence/adequacy and a free-text space for notes.

Dimension	Check (✓)	Notes/Revisions Needed
C: Constraints (word count, length, limits)	☐	
A: Audience (who is addressed, level of expertise)	☐	
S: Structure (organization, sequencing, clarity)	☐	
T: Tone (professional, supportive, persuasive, neutral)	☐	
R: Role (explicit identity given to the model)	☐	
O: Output format (list, essay, table, note, etc.)	☐	
F: Focus (Thematic priority of response)	☐	
F: Function (Communicativo purpoco or job of the output)	☐	

Overall Assessment

☐ Prompt is ready to use

☐ Prompt needs revision (specify): ______________________

7.7 Conclusion: From Prompt Repair to Prompt Literacy

This chapter has demonstrated that prompt engineering is not merely a matter of technical tweaking but a process of deliberate, iterative design. By applying CASTROFF as both a diagnostic tool and a generative framework, users learn to identify weaknesses in a prompt, articulate what is missing or ambiguous, and systematically revise it to suit a defined audience, tone, function, and communicative goal.

Through case studies and comparative outputs, readers have seen how seemingly minor revisions, such as specifying the AI's role or constraining the output format, can yield substantial improvements in clarity, relevance, and ethical alignment. The chapter has also illustrated the value of prompt audits and peer feedback in collaborative contexts, where shared language and consistency are critical.

By this point, readers should not only be able to repair weak or ambiguous prompts but also redesign them creatively and contextually for diverse applications. Prompt design has evolved from an invisible practice into a teachable and assessable discipline, with profound implications for education, professional communication, and the responsible integration of AI. CASTROFF is more than a checklist; it offers a conceptual lens for aligning human intent with machine responsiveness, supporting the broader goal of prompt literacy.

7.8 Self-Assessment Quiz

1. Which of the following revisions most clearly applies all eight CASTROFF elements to improve the prompt "Explain artificial intelligence"?

   a) Explain artificial intelligence in simple terms.

b) Act as a data scientist and explain artificial intelligence to beginners.

c) Act as a computer science teacher. In under 150 words, explain artificial intelligence to a group of 15-year-old students in a warm and engaging tone, using a real-life analogy. The output should be a numbered list suitable for inclusion in a classroom handout.

d) Explain artificial intelligence using no jargon and keep it short.

2. What is the primary function of a prompt audit in a collaborative setting?

   a) To score the quality of outputs

   b) To compare AI models for performance benchmarks

   c) To identify and revise flaws in prompt design using shared criteria

   d) To decide which team member writes the best prompt

3. In Case Study 2, which CASTROFF elements were most critical in transforming the vague climate report prompt into a usable government summary?

   a) Audience, tone, and role

   b) Constraints, focus, and output format

   c) Structure, function, and role

   d) Audience, focus, and structure

4. Why is defining the AI's role (the "R" in CASTROFF) particularly important when redesigning prompts?

   a) It makes the prompt more creative.

   b) It tells the AI what type of language or expertise to simulate.

   c) It reduces token usage in the output.

   d) It helps the AI generate multiple responses at once.

5. According to the chapter's conclusion, what is the broader value of CASTROFF beyond prompt repair?

   a) It automates prompt writing using templates.

   b) It teaches AI to evaluate its outputs.

   c) It supports human users in articulating and refining their intentions when designing prompts.

   d) It prevents hallucinations through rule-based formatting.

CHAPTER 8

Inclusive AI Prompting: Ethics, Accessibility, and Design for All

8.1 Prompting As Ethical Design

Prompt engineering is often mistaken for a neutral, technical activity. However, every prompt represents a series of decisions about audience, tone, framing, and purpose that shape how knowledge is retrieved, presented, and understood. These decisions do not simply guide the language model's behavior; they actively construct the epistemological boundaries of the interaction. In doing so, prompts determine which perspectives to include or exclude, amplify or omit certain voices, and shape how users perceive the system's authority. Prompting, therefore, is not only a functional task but a form of ethical design.

At the center of this claim is a recognition that prompts are never neutral. They reflect the assumptions, values, and worldviews of the humans who construct them. When left unexamined, these assumptions

H. Tavakoli, *Prompt Engineering for Everyone*,
https://doi.org/10.1007/979-8-8688-2338-1_8

can reinforce exclusion, particularly in domains such as education, healthcare, public service, and hiring, where AI-generated outputs carry material consequences. A prompt that defaults to dominant cultural or institutional perspectives may unintentionally erase alternative experiences or obscure forms of structural inequality. Ethical prompting does not guarantee the absence of bias, but it begins with the deliberate recognition that prompts shape outcomes, and those outcomes matter.

Despite growing awareness of AI's social risks, many users continue to treat effective prompting as a matter of fluency or productivity. Users often perceive a well-formulated prompt that yields coherent output as successful. However, fluency is not equivalent to fairness, and efficiency does not imply inclusivity. A prompt that produces polished language might still marginalize specific perspectives, reinforce stereotypes, or overlook emotional and cultural nuance. This issue becomes especially problematic when users mistake AI output for objective knowledge or institutional authority.

Recent research from professional bodies such as the Royal College of Psychiatrists and the American Psychiatric Association underscores the ethical implications of user instructions. These studies emphasize that prompt language, not just training data, can trigger the reproduction of social biases, diagnostic stereotyping, or exclusionary narratives. Where prompts fail to include lived experience, specify audience, or guide tone, the model defaults to statistically dominant representations. The results may not be malicious, but they are consequential: outputs that appear accurate may nevertheless be incomplete, insensitive, or misleading.

To address this, ethical prompting should be understood not as a matter of restraint but as a method of inclusion. We should not aim to restrict language or evade controversial topics but rather steer the model toward outputs that encompass a broader spectrum of human contexts, needs, and realities. Achieving this requires framing precise questions, identifying target users, defining tone and function, and acknowledging the power imbalances inherent in AI interactions. It is a communicative design discipline, one grounded in intentionality rather than neutrality.

Consider the prompt: "Explain autism." Although seemingly benign, it invites the model to produce a generalized, medicalized summary based on dominant narratives. A more inclusive alternative, *"Act as an autistic adult. Explain autism to teachers who want to create a safer classroom,"* reshapes the task entirely. This version introduces a role, identifies an audience, and clarifies the function. It centers lived experience and purpose over abstraction. The subject remains the same, but the ethical posture has changed.

Similarly, a prompt like *"List major inventions that changed the world"* tends to generate Eurocentric or industrial-age examples by default. The reframed prompt, *"List major inventions from African, Indigenous, and Asian civilizations that shaped world history,"* illustrates how design choices influence historical framing. This approach achieves inclusion not by censoring or correcting but by ensuring structural clarity and intentional scope.

These are not isolated fixes. They exemplify the ethical potential of prompt engineering as a reflective practice. The prompt determines the model's response, influences users' interpretation of that output, and defines whose perspectives appear in the exchange. To overlook these dynamics is to treat AI as a neutral mirror, one that reflects reality without bias. Engaging with them involves participating in the design of digital knowledge.

The CASTROFF framework operationalizes these principles by embedding ethical considerations into each prompt dimension. "Audience" prompts designers to account for excluded or underserved users. "Tone" helps avoid condescension, clinical detachment, or alienating informality. "Role" enables prompts to adopt situated perspectives, such as those of an activist, educator, or caregiver, rather than defaulting to abstract neutrality. "Focus" and "function" ensure that the goals of the interaction are grounded in real-world needs, not algorithmic convenience. Together, these elements make prompting not only more effective but also more accountable.

Inclusion through prompting is not an optional enhancement. It is a professional responsibility. As AI systems become embedded in public life, the consequences of poorly designed prompts are no longer limited to inconvenience; they include informational harm, cultural erasure, and exclusion from essential services. Ethical prompting mitigates these risks by foregrounding human context, social equity, and design transparency.

Viewed in this way, prompting becomes a practice of epistemic care. It invites us to ask: Whose knowledge counts? Whose experience is being centered? What assumptions are we encoding, and who might they leave behind? These questions do not slow down innovation; they make it more just.

8.2 The Myth of Neutral Prompts

A widespread misconception in AI interaction is that prompts can be neutral, that if phrased plainly or left open-ended, they will yield objective or universal results. In reality, prompts that appear neutral often operate as vessels for dominant social norms. They leave critical design decisions unspoken: Who is the imagined user? What kind of knowledge is valued? Which experiences are seen as typical or credible? Without clear guidance, the model defaults to the statistical majority within its training data, replicating existing biases under the guise of impartiality.

Language models are fundamentally probabilistic systems. They generate output based on patterns found in vast corpora of human language, much of it drawn from news media, internet forums, encyclopedic texts, and public archives. These sources reflect prevailing cultural attitudes, institutional norms, and demographic imbalances. The most likely answer is not necessarily the most inclusive, accurate, or just. In high-stakes contexts, this statistical normativity can become a form of epistemic harm.

Prompts that omit audience, tone, or context inherently create an ethical risk. For example, a prompt that asks, *"What makes a good leader?"* may produce an answer dominated by Euro-American business traits, such as assertiveness or decisiveness, unless specified otherwise. A question like "Describe successful parenting" risks returning heteronormative, nuclear-family ideals. These are not algorithmic failures; they are the result of human omission, choices not made, perspectives not specified, and constraints not articulated.

This bias-by-default is particularly dangerous in professional, public, or institutional environments. An AI summarizing legal rights, educational support, or health recommendations may reinforce norms that systematically underrepresent marginalized populations. For instance, a prompt such as *"Summarize mental health resources for teenagers" may* prioritize services familiar to urban, white, English-speaking users, excluding Indigenous practices, multilingual needs, or trauma-informed care models relevant to specific groups.

The idea that neutrality is achieved by saying less, or by omitting markers of identity, context, or power, is itself a design fallacy. Ethical prompting requires the opposite: greater specificity, intentional framing, and explicit scope. A so-called "neutral" prompt is not free of values; it is full of unexamined defaults.

To illustrate this, consider a prompt like "Explain climate change." This instruction is so broad that the model may produce an output based on mass-media consensus, which is often generalized and decontextualized. By contrast, a more carefully constructed prompt, *"Explain climate change from the perspective of a Pacific Islander community facing rising sea levels, using non-technical language for policymakers,"* clarifies role, audience, and function. The revised version does not introduce bias; it counters the illusion of neutrality by naming situated experience.

This shift from implicit to explicit design is not only more ethical; it is more effective. It produces outputs that are contextually grounded, communicatively purposeful, and socially responsive. The CASTROFF framework supports this shift by requiring designers to articulate their intent across multiple dimensions. "Constraints" prevent overgeneralization. The "audience" clarifies whose needs take priority, while "tone" controls emotional impact and suitability. "Role" establishes perspective, and "focus," along with "function," directs the model toward the intended outcome and purpose.

The model cannot determine these dimensions autonomously. It will, in the absence of guidance, adopt its most probable patterns, patterns that often exclude the very users who require the most careful design. In this light, neutrality is not a safeguard; it is a risk factor.

Inclusive prompting begins with recognizing that silence is not impartiality. It is a design decision with consequences. Prompts that fail to acknowledge the social world in which they operate risk amplifying its worst assumptions. Prompts that name their intentions, audiences, and responsibilities offer a path toward equitable AI interaction, not because they eliminate bias, but because they make its mitigation a conscious, designable task.

8.3 Recognizing Bias Traps in Prompting

Bias in prompting is rarely overt. More often, it operates through silence, through omissions, default assumptions, and decontextualized language that pass unnoticed. These subtle design flaws can significantly shape the model's output, reinforcing dominant perspectives while obscuring those of marginalized groups. In high-stakes contexts, such as healthcare, education, and public service, the consequences of this bias may not be immediately apparent at the surface level, but they are material and recurring.

The primary source of bias in prompt design lies not in overt prejudice but in the unexamined repetition of conventional frameworks. Designers create exclusionary outputs when they craft prompts that assume a generic user, disregard audience nuances, or mirror institutional jargon, not through deliberate harm, but because the AI system replicates dominant patterns from its training data. These patterns reflect historical asymmetries, linguistic centrality, and cultural dominance.

The CASTROFF framework offers a structured way to recognize and correct these patterns. When prompts leave critical dimensions, such as audience, tone, or function, unspecified, the model fills in the gaps with its defaults. These defaults often mirror elite, Western, able-bodied, and male-centric viewpoints.

Consider a prompt such as *"List three mental health tips for busy professionals."* Though seemingly innocuous, this prompt assumes a user who is employed, high-functioning, and economically secure. It prioritizes workplace productivity and neglects structural or trauma-informed perspectives. By contrast, a revised prompt, *"As a workplace therapist, offer three trauma-sensitive mental health strategies for junior staff, especially women and people of color working in public sector roles,"* specifies the speaker's role, names the audience, and embeds an ethical focus. It produces a more inclusive, grounded response, not by changing the topic, but by changing the prompt's perspective and social frame.

A similar pattern appears in prompts that omit the role entirely. For example, "Explain ADHD" typically yields a neutral, medicalized summary from the model's dominant sources. However, a prompt such as *"As an adult with ADHD, explain what you wish teachers understood about the condition"* introduces lived experience and a pedagogical function; the output shifts from generalization to specificity, from abstraction to situated knowledge.

Tone and function also play critical roles. A prompt that reads *"Provide information about domestic abuse resources"* may trigger a detached list of services. A more responsive version, *"Write a supportive message to someone experiencing domestic violence, with three actionable steps they can take today,"* introduces emotional safety, clarity of purpose, and a recognizable human voice. In this way, tone is not aesthetic; it is ethical.

Other common pitfalls include cultural generalizations and inaccessible formatting. A prompt like *"What are traditional wedding customs?"* is likely to produce a Eurocentric list unless explicitly reframed. A more inclusive version, *"Compare wedding traditions across Nigerian, Indian, and Peruvian cultures for an intercultural workshop,"* guides the model toward a broader representational range. Similarly, a dense, paragraph-style prompt may exclude users with cognitive disabilities or visual impairments. Designers can prevent unintended exclusion by structuring outputs as concise, screen-reader-friendly summaries.

Across these examples, a typical pattern emerges: prompts shape inclusion through the choices they make and the choices they omit. Designing inclusive prompts is not about policing language, but about recognizing that language structures access, trust, and recognition. The ethical risks arise when design decisions are made unconsciously or not made at all.

CASTROFF enables designers to bring those decisions into view. The audience clarifies who the prompt is for and whose needs are centered. The role positions the model as a representative of specific perspectives, rather than a default authority. Tone supports emotional appropriateness and relational trust. The output format ensures usability across diverse platforms and needs. Focus and function anchor the prompt to a defined communicative goal.

Ethical prompting is not a matter of perfection. No prompt can be entirely bias-free. Instead, ethical prompting demands a shift from default to design, from inherited assumptions to intentional construction. This practice begins with the awareness that prompting is not merely a technical exchange; it is a social act with consequences.

The following section examines how inclusive prompt design applies to accessibility and neurodiversity, one of AI interaction's most vital yet frequently neglected domains.

8.4 Designing for Accessibility and Neurodiversity

As AI systems become woven into the infrastructure of everyday life, accessibility in prompt design must be recognized not as a convenience, but as a core requirement of digital justice. Users do not approach language models from a shared baseline of cognitive capacity, linguistic fluency, or technological familiarity. Many users, whether neurodivergent, chronically ill, ageing, disabled, or navigating low-resource environments, interact with AI under conditions that remain structurally unacknowledged in mainstream design.

Conventional prompt engineering often presumes a "standard user": one who is non-disabled, mentally alert, literate in English, and familiar with digital interaction norms. The length, tone, formatting, and vocabulary of the prompt often embed this assumption implicitly rather than stating it directly. Such assumptions, when left unchecked, can exclude entire populations from meaningful participation in AI-mediated tasks. The exclusion is rarely intentional, yet it is predictably recurrent.

Neurodivergent users in particular, such as those with ADHD, autism, dyslexia, or Long COVID-related cognitive symptoms, may struggle with prompts that rely on abstract reasoning, dense language, or unstated expectations. Limited internet access or outdated devices further restrict users, often causing incompatibility with screen readers or excessive memory demands. Poorly designed prompts do more than confuse users; they actively communicate that the system excludes their needs.

Language models, by design, allow interaction in natural language. Many praise this affordance for enhancing accessibility. However, the benefits are conditional. Without careful and prompt design, even plain language can become exclusionary, especially when the tone is emotionally flat, the sentence structure is overly complex, or the prompt fails to define the model's role in supportive terms.

To design inclusively for diverse cognitive and physical needs, prompt construction must be attentive to structure, tone, and communicative scaffolding. For example, prompts that explicitly break tasks into short steps help users with executive functioning difficulties process requests incrementally. Direct, literal language reduces ambiguity for autistic users and others who prefer concrete instruction. Repetitive or simplified phrasing can support those experiencing brain fog, fatigue, or memory impairment. These are not special adaptations; they are inclusive defaults that enhance usability for all.

Consider a user navigating a healthcare system with limited digital literacy. A prompt asking, *"Describe your symptoms, medical history, and current medications,"* may overwhelm or confuse the user, especially if they are stressed or unwell. A more accessible prompt, *"Let us take this one step at a time. First, describe what is bothering you today. Then tell me about any past illnesses. We will talk about medications after that,"* creates a pacing structure that recognizes cognitive load and emotional state. The result is not just clearer information gathering, but a more humane interaction.

The CASTROFF framework reliably embeds accessibility at the design point. The "structure" element supports chunked, stepwise output that is easier to parse. "Tone" helps the prompt remain emotionally appropriate, avoiding overly clinical, robotic, or rushed language. "Output format" can be tailored for screen reader compatibility, printable layouts, or text-only presentation, reducing cognitive barriers to retrieval and reuse. "Audience" reminds designers that users are not abstractions; they are individuals with distinct capacities, constraints, and lived conditions.

Even when prompts appear to work well for general users, they may fail silently for others. A paragraph-heavy instruction might exclude a dyslexic reader. A culturally idiomatic phrase may confuse non-native speakers. A prompt that relies on metaphors or humor may be inaccessible to users who process language literally. In such cases, the design failure lies not in the AI's generative capacity, but in the human interface that shaped its input.

By contrast, an accessible prompt can restore agency. A simplified request may help an older adult complete a digital form independently. A role-specified instruction can support a neurodivergent worker preparing self-advocacy language for their employer. A scoped, calming query may reduce anxiety for someone facing an unfamiliar or intimidating bureaucratic system. These interactions are not minor; they are moments in which technological systems either extend or withhold participation.

Accessibility in prompting is also a matter of linguistic equity. For multilingual users, prompts should avoid contractions, idioms, or culturally specific metaphors. International formats for dates and numbers increase clarity. Prompts must account for the possibility of shared or unstable devices and design accordingly, avoiding unnecessary memory load, reducing response complexity, and ensuring fallback options for users who cannot easily rephrase or retry.

This is not an argument for simplification as a concession. It is a call for sophistication in ethical design. Inclusive prompting does not reduce the intelligence of the model; it broadens its communicative relevance. It aligns AI behavior with the lived realities of a pluralistic user base.

The implications of inaccessibility are not abstract. In a society increasingly mediated by automated interaction, inaccessible prompts can result in denied access to services, misunderstood rights, or increased vulnerability to misinformation. Public and private institutions deploying LLMs in frontline services must treat accessible prompt design as a core obligation, an infrastructural standard rather than an optional feature.

Prompt engineering, when approached through the lens of accessibility and neurodiversity, becomes a form of civic design. It upholds the principle that all users, regardless of cognitive style, disability status, or linguistic background, deserve systems that recognize and support their dignity.

In the next section, we examine how these principles scale to the design of public-facing systems, where prompt accessibility is not only a question of individual fairness but also a matter of democratic accountability.

8.5 Design for All Checklist

Use this list as a quick screen before finalizing prompts for public- or user-facing tasks:

- **Readability**: Aim for CEFR B1–B2 (short sentences, concrete vocabulary).
- **Multilingual fallbacks**: Provide plain-English defaults, along with options for translation or simplified terms.
- **Trauma-informed phrasing**: Use supportive, non-judgmental tone; avoid shaming or alarmist wording.
- **Consent and opt-out**: Signal when users may skip steps, withdraw, or choose not to share certain information.
- **Device constraints**: Design prompts for low bandwidth, small screens, and screen reader compatibility.
- **Tone adjustments**: Check cultural sensitivity; avoid idioms, jokes, or culturally specific metaphors.
- **Structure**: Break complex requests into clear, sequential steps.
- **Role clarity**: Frame the model as a helper or advisor, not as a gatekeeper or authority figure.
- **Output format**: Specify bullet points, short answers, or plain text to aid accessibility.
- **Fairness lens**: Ask, "Whose *needs are not served by this wording?"*

8.6 Inclusive Prompting in AI-Powered Public Services

The expansion of AI into public service infrastructure has transformed how individuals access essential rights and resources. Interactions that once occurred through trained human intermediaries, such as booking medical appointments, appealing government decisions, or requesting social support, are increasingly mediated by AI-powered interfaces, including chatbots, automated helpdesks, and natural language systems. These systems promise efficiency, but they also introduce new conditions for inclusion. For many users, access to public services now depends not only on eligibility or entitlement but on their ability to interpret, formulate, and respond to AI-generated prompts.

This shift raises profound questions about digital equity and civic participation. Public sector AI system designers frequently focus on a narrow user archetype, an individual who possesses digital literacy, cognitive fluency, emotional stability, and proficiency in the dominant language. While this profile may be statistically standard among early adopters of AI, it excludes a wide array of users, including older adults, disabled individuals, recent migrants, low-literacy communities, and those experiencing stress, trauma, or institutional mistrust.

The consequences of such exclusion are not theoretical; they are honest and tangible. A pensioner receiving a penalty notice may be unable to navigate a chatbot to contest it. An autistic teenager attempting to schedule a medical appointment may find the interface ambiguous or emotionally alienating. A person experiencing cognitive overload may give up mid-process when asked to complete a form framed in legalistic language. In each case, the obstacle is not user competence, but the system's failure to accommodate real-world variation in capacity, context, and need.

In this context, prompting functions as infrastructure, requiring designers to apply the same rigor, inclusivity, and foresight as they would to physical accessibility measures. Just as buildings require ramps and signage to ensure equitable access, digital systems must incorporate prompt structures that reduce cognitive, linguistic, and emotional barriers.

The CASTROFF framework offers a systematic approach to fulfilling this obligation. Each of its components offers a point of intervention for designers and institutional stakeholders seeking to make AI-powered public services more inclusive.

"Constraints" guide prompt writers to reduce complexity by limiting memory demands, technical vocabulary, or the number of steps. A public service prompt that asks users to "list all supporting documents with relevant legal references" assumes legal knowledge and organizational clarity. Reframing this as "Tell us which documents you have. We will explain what else might be needed" reduces pressure while signaling support.

The "audience" element prompts designers to consider who is likely to interact with the system under unequal or constrained conditions. Refugees completing immigration forms, first-time applicants for social housing, or individuals navigating disability claims may lack prior exposure to bureaucratic language. Identifying these users explicitly helps align the prompt tone and structure with their actual circumstances.

"Tone" plays a critical role in cultivating trust and emotional safety. Public-facing AI systems should avoid legalistic phrasing, performative neutrality, or abrupt commands. A tone that is calm, encouraging, and forgiving of uncertainty can significantly improve the user experience, especially for those interacting with systems under stress.

"Role" enables AI systems to adopt identities that signal care, solidarity, or guidance, rather than institutional surveillance. For example, an AI system responding as a "community benefits advisor" rather than a generic "assistant" changes the interaction dynamic. It positions the system as a helper rather than an auditor.

The "output format" ensures that responses are readable across various devices and conditions. In public contexts, users may be on mobile data, using shared devices, or relying on assistive technologies. Prompts that specify short, bullet-pointed answers, screen reader compatibility, or multilingual summaries accommodate these conditions and promote equitable access.

"Focus" anchors the prompt to what the user needs to accomplish most, avoiding drift into irrelevant details. For example, a prompt that says, "Explain your change in financial status," may generate confusion. A focused version, "Tell us if your income has changed since your last claim. This helps us adjust your benefits fairly", clarifies both purpose and ethical framing.

Finally, "function" defines the type of exchange that is taking place: Is the prompt gathering information, offering reassurance, facilitating an appeal, or confirming legal compliance? Without such clarity, users may hesitate to respond or misinterpret the stakes of their interaction.

Together, these elements transform prompting from a user challenge into a site of public service. They allow public sector designers to anticipate exclusion, embed clarity, and build systems that reflect civic values.

This shift also requires a new mindset about public technology. Writers must design prompts anticipating that users face non-ideal conditions, treating factors like fatigue, distrust, multitasking, shared devices, poor connectivity, and emotional distress as common scenarios rather than exceptions. They are the lived realities of many citizens engaging with digital governance.

Consider two versions of a benefits prompt. One asks, *"List your National Insurance number, address, and grounds for appeal."* This version treats the interaction as purely transactional. A reframed version reads: *"Please enter your National Insurance number and address. Then, in your own words, explain why you are appealing. You may skip any question and return later. If unsure, share what you can; we will guide you from there."* The content remains similar, but the framing transforms the system from a gatekeeper into a support structure.

Prompting, in the context of AI-powered public services, is not a technical detail. It is a question of democratic design. If prompts are inaccessible, unclear, or emotionally alienating, they reduce public systems to mere machinery that serves only the privileged. Inclusive prompting, by contrast, affirms that public institutions remain accountable to all, not only to those fluent in their language and logic.

In the next section, we explore how CASTROFF can be extended beyond individual design choices into systemic tools for evaluation, audit, and institutional governance, ensuring that inclusion is not left to chance but built into the infrastructure of prompt engineering itself.

8.7 Inclusive Prompt Rewrites: Positive vs. Negative Examples

The following examples illustrate how small design choices shift prompts from exclusionary to inclusive. A short rationale follows each pair.

1. **Benefits Intake Form**
    - **Negative**: "List all supporting documents with relevant legal references."
    - **Positive**: "Tell us which documents you already have. We will explain what else might be needed."
    - **Rationale**: The positive version reduces cognitive load, avoids assuming legal literacy, and frames the system as supportive rather than punitive.
2. **Appointment Reminder**
    - **Negative**: "Your appointment is scheduled for 14:30. Failure to attend will result in cancellation."

- **Positive**: "Your appointment is booked for 2:30 pm. If you cannot attend, you can reschedule or cancel without penalty."
- **Rationale**: The positive version uses plain time formats, avoids threatening language, and clarifies user options.

3. **Debt Letter**
   - **Negative**: "You are in arrears. Immediate repayment is required to avoid further action."
   - **Positive**: "Our records show an unpaid balance. You can set up a payment plan, request support, or contact us if you believe this is incorrect."
   - **Rationale**: The positive version acknowledges user dignity, offers options, and frames the system as collaborative rather than adversarial.

8.8 CASTROFF for Ethical and Inclusive Prompting

The CASTROFF framework, introduced in earlier chapters as a professional standard for prompt engineering, attains its most whole ethical significance in the context of inclusion. While its components, constraints, audience, structure, tone, role, output format, focus, and function, each serve as practical controls for shaping AI outputs, together they form a model for sustained accountability. CASTROFF enables prompt designers, institutions, and public systems to move beyond individual intuition toward a repeatable, transparent method of inclusive practice.

Inclusion in prompting is not an act of goodwill. It is a form of design governance, an obligation to consider the full range of human needs, experiences, and risks that arise in AI-mediated interaction. Prompting decisions shape who can participate, who is recognized, and who benefits from AI. Without tools to formalize inclusive decision-making, institutions risk embedding exclusion into automated workflows under the illusion of neutrality.

CASTROFF is structured to prevent such drift. By requiring explicit articulation of audience, tone, and communicative function, it forces prompt designers to name the users they serve and the outcomes they intend. This practice transforms prompting from a convenience-based interaction into a values-aligned exchange, one that foregrounds social responsibility as a design principle.

Consider, for example, a team developing AI-based intake forms for a social benefits system. Using CASTROFF, they begin by identifying constraints such as limited internet access and low digital literacy. They then define the audience as first-time applicants, including recent migrants and users with disabilities. The structure is adjusted to feature step-by-step prompts with visual spacing. The tone is revised to be warm and supportive, rather than procedural. The role is set as a "community advisor," not an impersonal interface. The output format is made compatible with mobile devices and screen readers. The focus becomes user understanding and clarity of eligibility. The function is clearly stated as supporting application completion, not legal verification. In this workflow, inclusion is not an afterthought; it is integral to prompt architecture from the start.

Beyond individual prompts, CASTROFF also supports institutional *prompting audits*. Teams can evaluate their prompt templates by reviewing whether each dimension is clearly defined and ethically coherent. Are prompts structured for cognitive access? Do they name marginalized audiences explicitly? Does the tone convey trust and emotional safety?

Are functions aligned with user intentions rather than administrative efficiency? These questions, grounded in CASTROFF, provide a blueprint for collective review, peer accountability, and continuous improvement.

This model of prompt auditing mirrors established practices in fields such as content design, accessibility evaluation, and algorithmic fairness. Each discipline developed governance frameworks after recognizing that addressing bias requires systemic solutions, documented standards, and shared evaluative tools, rather than relying solely on good intentions. The CASTROFF framework establishes these standards specifically for prompt design.

Inclusive prompting also demands temporal awareness. A prompt that is accessible to a user on one day, when they are well-rested, supported, and confident, may be exclusionary under different conditions, such as moments of grief, illness, or stress. The CASTROFF framework's "focus" and "tone" dimensions specifically enable designers to accommodate these fluctuations, redefining inclusion as contextual adaptability rather than mere demographic representation, a design approach that responds to the inherent unpredictability of human needs.

Furthermore, CASTROFF provides a language for interdisciplinary collaboration. Accessibility experts, frontline workers, ethics reviewers, and legal teams can utilize the framework to engage in prompt design discussions with clarity and a shared vocabulary. It bridges technical implementation with human-centered evaluation, allowing diverse stakeholders to surface risks, recommend improvements, and defend design choices with reasoned transparency.

When implemented systematically, CASTROFF fosters ethical maturity across the prompt lifecycle. It allows institutions to build *equity into automation*, not only by correcting exclusion after it occurs but by preventing it through design foresight. It positions prompt engineering not as a creative improvisation, but as a formalized literacy, a civic competency as essential to digital governance as legal plain language or architectural accessibility.

In this view, inclusive prompting is not merely the responsibility of designers, but of organizations, governments, and educational systems that deploy AI in public life. CASTROFF provides these actors with a framework to act responsibly, audit rigorously, and build trust through transparent and inclusive digital language.

The final section shifts from demographic inclusion to *situational inclusivity*, expanding the ethical scope of prompting to encompass emotional states, cognitive fluctuations, and the broader politics of human vulnerability.

8.9 Reframing Inclusion Beyond Demographics

Much of the discourse on inclusive prompting centers on demographic categories, including race, disability, neurodivergence, language, and socioeconomic status. This framing is essential, but it is incomplete. Ethical prompt design must also account for *contextual exclusion*, the temporary, situational, or affective conditions that shape a user's interaction with AI. Inclusion is not only about who the user is but also about when and where they are engaging with the system.

A prompt that is accessible to a confident, alert user may fail for that same individual when they are grieving, overwhelmed, or cognitively exhausted. Likewise, a user may temporarily lack access to stable internet, private space, or the emotional regulation required to parse complex instructions. In these moments, the barrier is not a fixed identity trait, but a shifting condition of vulnerability.

This insight complicates the task of inclusive prompting. It challenges the assumption that user characteristics are static or easily mapped to design templates. Instead, it requires prompt designers to build situational empathy into their workflows, acknowledging that inclusion must be flexible, responsive, and adaptable to the fluctuating capacities of humans.

Consider an example from a healthcare setting. A prompt that reads, *"List your symptoms, medications, and prior diagnoses in chronological order,"* may function well for a clinician but could overwhelm a patient who is in pain, anxious, or cognitively fatigued. A restructured prompt, *"Let us take this step by step. First, tell me what is bothering you today. Then we can go over your past treatments together,"* not only lowers cognitive load but signals care. The difference lies not in the information requested, but in the tempo, tone, and pacing of the interaction.

This reorientation also helps avoid a common misstep in inclusive design: the tendency to over-categorize or over-medicalize users. Not every user with difficulty concentrating requires a clinical label. Not every emotional barrier is evidence of a diagnosis. Designing for fluctuating states, rather than fixed traits, respects users as complex, dynamic individuals whose needs cannot be fully anticipated or pre-classified.

The CASTROFF framework supports this flexibility by embedding adaptability across its components. Designers can adjust the tone to communicate either calmness or urgency based on the user's probable state. They may modify the structure to provide either step-by-step guidance or open-ended scaffolding. Furthermore, they can refine the function to prioritize emotional support alongside task accomplishment. Crucially, role allows the model to inhabit supportive stances, serving as a guide, caregiver, or peer, rather than impersonal authority figures that may intimidate or alienate users in distress.

Contextual inclusion also matters in high-stakes environments. In legal or financial settings, users may interact with AI while experiencing housing insecurity, language barriers, or institutional distrust. A prompt that assumes comfort with legal jargon or familiarity with procedural norms will fail to serve these individuals, even if they do not belong to a traditionally marginalized group. Thus, inclusive prompting must move beyond demographic risk profiling and toward situational *anticipation*.

This perspective has implications for both the design and evaluation of prompts. It urges designers to ask: How might this prompt perform under conditions of stress, distraction, or low comprehension? Is it structured in a way that allows partial engagement, repetition, or recovery from confusion? Does it offer the user emotional permission to be uncertain, slow, or overwhelmed?

Prompts that address these questions do more than meet compliance goals; they enhance the resilience of human–AI interaction. They create systems that bend with the user's needs rather than breaking under the weight of their variability.

Inclusion, in this expanded sense, is not a checklist. It is a commitment to building AI systems that respond to the full complexity of human life, not only in terms of identity but in terms of context, capacity, and care.

8.10 Chapter Summary

This chapter has reframed prompt engineering as an ethical and civic design practice, particularly in contexts where artificial intelligence intersects with public services, human vulnerability, and institutional systems. The framework fundamentally acknowledges that *prompts* inherently carry bias; they embed assumptions about users' identities, appropriate knowledge presentation formats, and which experiences qualify as legitimate or relevant.

We began by establishing that every prompt carries social consequences, not only in what it says but in what it fails to ask. Functional fluency or model coherence is not a sufficient benchmark for success. Effective prompts must also be inclusive, transparent, and deliberately constructed to counteract the probabilistic biases of large language models.

The chapter challenged the myth of neutrality, demonstrating how omissions in tone, audience, or function can inherently produce biased outputs. Through real-world examples, we examined how exclusion is reproduced not just through data or algorithms, but through the human decisions embedded in prompt design. We explored six common bias traps and demonstrated how the CASTROFF framework enables the systematic diagnosis and repair of these biases.

We then expanded the concept of inclusion to encompass accessibility and neurodiversity. The CASTROFF framework developed inclusive prompting to address cognitive overload, emotional distress, and design barriers that impact diverse users across various contexts. Here, we argued that accessibility is not a special feature but a baseline responsibility in digital systems that serve the public.

Researchers identified public sector AI applications as a critical priority for implementing ethical prompting practices. In this context, prompts become civic infrastructure, structuring how citizens access healthcare, appeal decisions, and navigate legal obligations. The chapter emphasized that prompting in such domains must be designed for real-world conditions, including limited literacy, unstable connectivity, emotional volatility, and institutional mistrust.

Finally, we extended the CASTROFF framework into two crucial areas. First, we demonstrated how it can serve as an *institutional audit tool*, facilitating prompt literacy across design teams, accessibility reviewers, and ethical oversight bodies. Second, we redefined inclusion beyond demographic categories, arguing that ethical prompting must also account for *contextual and situational vulnerability*. By anticipating shifts in cognitive capacity, emotional state, or environmental constraint, prompt designers create systems that are more resilient, more just, and more human-centered.

In this view, prompting is not merely about instructing machines; it is also about guiding them. It is about shaping systems that reflect our collective priorities, uphold our ethical commitments, and widen the boundaries of digital participation.

8.11 Glossary of Key Terms

- **Bias**: A repeated preference or assumption that leaves out certain people or views
- **Harm**: Negative effects on people, such as unfair treatment, exclusion, or stress
- **Protected characteristics**: Aspects of identity, such as age, race, gender, disability, or religion, that laws protect from discrimination
- **Accessibility needs:** Adjustments that ensure information remains usable for individuals with diverse abilities or conditions
- **Neutral prompt**: A prompt that looks objective but often hides cultural or social assumptions
- **Trauma-informed**: Language that avoids blame or harsh tone, showing care for people who may have lived through difficult events
- **Consent**: Clear permission from the user to share or withhold information
- **Opt-out**: A choice that lets the user stop or skip a step without punishment
- **Inclusive design**: Creating prompts so they work for as many people as possible, across different situations
- **Cultural awareness**: Sensitivity to how words, tone, or examples may carry different meanings for different groups

8.12 Self-Assessment Quiz

1. Why is it problematic to leave the audience unspecified in a prompt?

   a) It increases the risk of computational inefficiency.

   b) It limits the model's ability to vary tone.

   c) It causes the model to default to dominant, often biased, assumptions.

   d) It produces longer outputs than necessary.

2. Which of the following best describes the myth of neutrality in prompting?

   a) Prompts with neutral wording are always safer.

   b) Prompts that lack emotional tone perform better.

   c) "Neutral" prompts often mask unacknowledged cultural assumptions.

   d) Neutrality should always be the goal of inclusive prompt design.

3. What makes a prompt inclusive for neurodivergent users?

   a) The use of academic vocabulary and general instructions

   b) Minimal structure to encourage open-ended thinking

   c) Role-based phrasing and abstract metaphors

   d) Clear steps, literal language, and emotional tone appropriate to the context

4. How does CASTROFF help institutional teams assess and promote inclusive practices?

   a) By generating alternative prompts automatically

   b) By applying a consistent, auditable structure across design elements

   c) By prioritizing performance over ethical concerns

   d) By simplifying legal compliance for automation workflows

5. Why must inclusive prompting extend beyond demographic categories?

   a) Most biases occur at the system level, not the user level.

   b) Situational vulnerability, such as grief or fatigue, can affect any user.

   c) Fixed categories are legally protected, but temporary states are not.

   d) Language models lack training to recognize demographic identity.

CHAPTER 9

Prompt Debugging and Safety in Sensitive Contexts

9.1 When Prompt Mistakes Become High-Risk

In many professional contexts, prompt engineering is still viewed as a matter of efficiency, a way to extract quicker, clearer, or more accurate responses from large language models (LLMs). However, when prompting occurs in sensitive domains such as healthcare, education, criminal justice, or employment, the consequences of poorly constructed prompts are no longer theoretical or cosmetic; they become a matter of real-world impact. Similar risks also arise in technical fields, including engineering, cybersecurity, and data governance. In these settings, imprecision, omission, or ambiguity in the prompt can directly compromise human safety, equity, and institutional trust.

Organizations increasingly deploy LLMs in areas that shape critical decisions, diagnose symptoms, provide legal overviews, evaluate job applicants, and deliver feedback to students. These systems are often used not by technical experts, but by clinicians, educators, administrators, or

H. Tavakoli, *Prompt Engineering for Everyone*,
https://doi.org/10.1007/979-8-8688-2338-1_9

members of the public, many of whom assume that the model's fluency signals correctness. When the prompt guiding these interactions is vague or incomplete, the system may generate hallucinated facts, replicate systemic biases, or respond in ways that are emotionally or culturally inappropriate. The risk extends beyond output quality to encompass social harm.

In the medical domain, for instance, prompt design becomes a matter of clinical safety. Consider a user experiencing slurred speech and arm weakness who enters a general question such as *"Why is my speech slurred and my arm weak?"* A model lacking a well-structured prompt may default to benign explanations, such as stress, dehydration, or fatigue, while overlooking signs of a stroke. Without explicit constraints, urgency signals, or escalation pathways built into the prompt, the model is unlikely to issue a critical safety recommendation. Stanford's Institute for Human-Centered Artificial Intelligence (2023) found that LLMs frequently provided incomplete or outdated explanations of cancer treatment options unless users imposed explicit constraints on source accuracy, scope, and tone. Similarly, the UK National Health Service reported that several AI-based health chatbots failed to identify symptoms of stroke and sepsis due to vague prompts and limited contextual awareness. In these cases, a lack of prompt precision becomes a vector for harm, not a technical flaw, but a clinical liability.

Legal applications introduce a different, but equally dangerous, set of risks. Prompts that guide LLMs in reviewing court documents, summarizing prior arrests, or suggesting sentencing guidelines can unconsciously reproduce entrenched biases. A prompt such as *"Summarize this defendant's criminal history"*, without qualifiers for context, expungements, or ethical framing, may yield responses that amplify racial disparities embedded in the training data. A 2021 analysis in Science, building on findings from ProPublica and academic researchers, showed that LLMs frequently reflected the statistical overrepresentation of marginalized groups in judicial records. When prompts fail to specify fairness constraints or legal nuance, they become complicit in perpetuating injustice.

In education, AI-generated feedback is transforming how students engage with writing, critical thinking, and revision. However, prompts that lack sensitivity to learner context, such as *"Critique this student's writing,"* can lead to responses that are overly harsh, excessively formal, or cognitively misaligned. For neurodivergent learners or students with prior trauma, such feedback may feel alienating or demoralizing, even when technically accurate. Effective debugging in educational contexts requires more than clarity; it must also consider tone, audience, and the intended emotional and pedagogical impact.

Employment settings offer further complexity. Recruiters who use prompts to assess résumés or simulate interview responses may introduce algorithmic bias through vague or decontextualized language. A question like *"Rate this applicant's leadership ability"* assumes a shared cultural understanding of what leadership looks like, often privileging extroverted or assertive styles. Without inclusive definitions or role framing, prompts risk penalizing applicants who demonstrate leadership through collaboration, mentorship, or resilience, especially those from underrepresented or non-dominant cultural backgrounds.

Across these domains, the pattern is clear: the quality of the prompt governs the safety of the output. The risks are not limited to technical malfunctions. They are structural, embedded in the design of the prompt itself. An LLM does not pause to clarify vague instructions or question its assumptions. It proceeds fluently, using the most statistically probable completions from its training data. An underspecified prompt causes the system to default to its most frequently observed patterns, which rarely remain neutral. They reflect historical inequities, cultural biases, and normative assumptions baked into the data. In high-stakes settings, this behavior transforms prompting into an ethical act.

Prompt debugging, therefore, is not a discretionary skill. It is a necessary component of responsible AI use in sensitive contexts. It requires the user to take deliberate control over how the model interprets its task, ensuring that the output aligns with human intention, ethical boundaries, and contextual appropriateness. This chapter equips readers to do just that: not only to identify and correct prompt flaws but to anticipate where harm might arise and how to design prompts that protect against it.

9.2 What Occurs When Users Fail to Specify Prompts?

Large language models operate by identifying and reproducing statistical patterns in language. When a prompt lacks critical details, such as the intended audience, the AI's assumed role, the expected tone, or any ethical constraints, the model does not pause to request clarification. It proceeds as though the instructions were complete, confidently generating a response based on what is most common or dominant in its training data.

This default behavior carries profound implications, particularly in sensitive domains. When users fail to define the frame of the interaction, the system implicitly substitutes its assumptions. These assumptions are rarely neutral. These patterns disproportionately represent certain cultural norms, knowledge domains, and communication styles, thereby shaping the system's outputs. Without user-defined constraints, the model tends to reproduce Western, Anglophone perspectives, adopt emotionally detached or clinical tones, and apply generalizations that may exclude or misrepresent marginalized populations.

To understand the dynamics of these omissions, we can draw a metaphor from the concept of coercive control in psychology and sociology. Coercive control does not always rely on force; rather, it manifests when one party silently takes over decision-making in the absence of resistance or clearly defined boundaries. Similarly, when a user submits a vague prompt, the model assumes authorship quietly. It writes not according to the user's values or intentions, but according to probabilistic defaults.

For instance, a prompt that does not specify the audience may result in a response pitched to a generic adult reader with high literacy and technical proficiency. Failing to address tone may lead the model to adopt a formal or impersonal style, which users may perceive as dismissive or emotionally cold. These decisions, although statistically justifiable from the model's perspective, can lead to significant misalignments in healthcare, education, employment, and justice domains, where nuance, empathy, and contextual sensitivity are crucial.

What makes this risk especially dangerous is its invisibility. The model does not signal that it is making assumptions. It does not indicate which gaps it has filled or on what basis. Instead, it proceeds with linguistic fluency that may mislead users into believing the output is accurate, appropriate, or aligned with their intent. Hidden assumptions in practice can introduce bias, perpetuate exclusion, or compromise the user's original intent, making the CASTROFF framework indispensable. By explicitly defining constraints, audience, structure, tone, role, output format, focus, and function, users reclaim authorship over the interaction. Prompt design becomes an act of deliberate, structured expression rather than an open-ended query that leaves too much to chance.

In high-stakes contexts, underspecified prompting is not a minor stylistic flaw; it is a systemic risk. Every omission in a prompt creates an opening for the model to insert assumptions. Without constraint, it writes from the center of the statistical map, not from the margins where nuance and ethical care are often required. Designing prompts with explicit intent is, therefore, not just a technical skill; it is an ethical responsibility.

In the next section, we explore specific debugging techniques that allow users to identify and correct such risks before they cause harm.

9.3 Practical Techniques for Debugging Prompts

Prompt debugging is not a supplementary step in prompt engineering. It is a core design responsibility, particularly in sensitive domains where the stakes of failure include harm, exclusion, or systemic bias. While it is tempting to view prompt generation as a one-time creative act, the reality is that effective prompts emerge through iteration, testing, and diagnosis. Debugging is a process by which users regain control over what the model says, how it expresses itself, and to whom it is speaking.

This section introduces five professional techniques for debugging prompts. Each reflects the central insight of the CASTROFF framework: that clarity, constraint, and audience-awareness are essential for safe and accountable interaction with large language models (LLMs). Each subsection illustrates how a loosely defined prompt can be re-expressed in clearer CASTROFF terms to reduce hidden assumptions.

1. **Iterative Variation for Output Stability**

   A well-formed prompt should produce consistent, context-appropriate outputs across reasonable rephrasing. One of the simplest techniques for debugging a prompt is to rerun it with minor linguistic variations, changing tone, phrasing, or structure, while preserving the core instruction. If the model's outputs shift unpredictably or diverge significantly in tone or accuracy, the prompt may be underspecified or sensitive to surface cues.

   A model may produce different summaries of a medical report for a patient depending on whether the user phrases the prompt formally or conversationally. An unstable prompt in this context suggests the need for more explicit control over tone, audience, or function.

2. **Reverse Application of the CASTROFF Framework**

   CASTROFF is not only a design framework, but it is also a powerful diagnostic tool. When a prompt produces problematic or unpredictable output, reverse analysis can help pinpoint which elements were left vague. The user may ask

   - Was the intended audience defined clearly?
   - Did the prompt specify an appropriate tone or role?
   - Was the output format structured for accessibility or reuse?

   This method reveals implicit assumptions that the model makes in the absence of user guidance. The framework also identifies where human intervention must reintroduce ethical, linguistic, or contextual constraints to restore authorship.

3. **Edge-Case Persona Testing**

   Researchers should test prompts with *edge-case personas*: hypothetical users whose needs, contexts, or identities deviate from the statistical norm, including individuals with low literacy, cognitive fatigue, non-dominant cultural frameworks, or emotional vulnerability.

   For example, a prompt designed to deliver financial advice should be evaluated for clarity, accessibility, and emotional tone when interpreted by a first-time user, a stressed caregiver, or someone unfamiliar with financial jargon. A prompt that falters with edge-case personas risks *exclusion, confusion, or harm*, necessitating restructuring to accommodate a broader spectrum of real-world users.

4. **Structural Decomposition and Justification**

   Debugging also involves looking inside the model's output. One way to do this is to ask the model to break down its response into labelled components, such as definitions, justifications, or sequential steps. A model that can explain why it generated a particular response provides greater transparency, enabling users to identify breakdowns in logic, relevance, or fairness.

   This technique not only surfaces weak internal reasoning but also helps users assess whether the output structure aligns with the intended communicative function. In professional or public service settings, such decomposition often becomes a prerequisite for trust and traceability.

5. **Secondary Critique and Self-Audit Prompting**

   Finally, users can instruct the model to *critique its output* by submitting a follow-up prompt that assigns it a reviewer's role, such as

   - "Evaluate the previous response for emotional appropriateness."
   - "Identify any assumptions or omissions that might affect a vulnerable user."

This meta-level prompting transforms the model from a passive respondent to an active evaluator. It often reveals *unacknowledged bias, misaligned tone, or contextual errors* that the original prompt may have overlooked. In regulated or institutional environments, such self-audits can serve as a lightweight form of ethical review before deployment.

These techniques are not mutually exclusive. The most robust prompt debugging processes combine multiple strategies, testing for structural reliability, ethical alignment, tone appropriateness, and contextual inclusivity. Each method reflects the deeper insight that prompt design is not just instruction; it is risk *shaping*. What is left vague in a prompt will be filled in by the model's defaults, and those defaults are not neutral.

In the next section, we examine how these debugging techniques operate in real-world settings, across healthcare, education, justice, and employment, where the cost of unexamined assumptions is particularly high.

9.4 Prompt Debug Protocol

To support repeatable practice, organizations can formalize prompt debugging into a structured protocol. This process helps practitioners move from intuition toward systematic diagnosis and correction. The protocol unfolds in six steps:

1. **Reproduce:** Confirm the issue by running the prompt multiple times, documenting the variation in outputs. Reproducibility establishes whether the fault is systemic or incidental.

2. **Isolate:** Identify which CASTROFF dimension is most likely responsible (constraint, audience, structure, tone, role, output format, focus or function). For example, if outputs drift in length, the "focus" or "output" dimension is at fault.

3. **Hypothesize:** Formulate a clear explanation of why the prompt is failing. This step turns vague dissatisfaction into a testable claim (for instance, "The ambiguity lies in scope definition.")

4. **Patch:** Adjust the prompt in a targeted way, addressing the suspected weakness. Avoid wholesale rewriting; the goal is incremental refinement so that causal links remain visible.

5. **Re-test:** Run the revised prompt with both typical and adversarial cases, inputs designed to stress the boundaries of the instruction. This stage checks whether the patch resolves the issue without introducing new failures.

6. **Document:** Record the fault, hypothesis, patch, and result. Documentation strengthens accountability, enables team learning, and guarantees that improvements are reproducible across different contexts.

By embedding this protocol into professional practice, prompt debugging becomes less about improvisation and more about disciplined engineering. The cycle of reproduce, isolate, hypothesize, patch, re-test, and document establishes a reliable method for refining prompts in sensitive and high-stakes environments. This text-based sequence also acts as a simple escalation pathway, indicating when a prompt may require a human review before deployment.

9.5 Case Studies in Prompt Debugging

To understand the real-world impact of prompt debugging, it is helpful to examine how flawed prompts behave in applied settings and how careful revision can prevent harm. This section presents five illustrative case studies from healthcare, law, education, employment, and customer service. Each case traces a breakdown in prompt design, followed by a correction using CASTROFF-based techniques. Together, they demonstrate that debugging is not simply about improving clarity; it is about restoring human intent, preventing systemic bias, and ensuring contextual safety.

9.5.1 Case 1: Medical Triage—Structuring for Urgency

A healthcare organization deployed a chatbot to respond to patient queries, including one prompt that stated

> *"Give possible reasons why someone might feel short of breath."*

The model generated general, low-risk explanations, including anxiety, lack of fitness, or exposure to cold air. However, the prompt omitted potentially critical causes such as pulmonary embolism or heart failure. The issue lay not in factual error, but in the *absence of clinical framing.*

The debugged prompt introduced structured triage logic:

> *"Act as a clinical triage assistant. Organize possible causes of shortness of breath into emergency, moderate, and mild categories. Use plain language, and advise the user to seek urgent care if symptoms fall into the emergency category."*

This revision clarified the model's role, added a safety failsafe, structured the output, and foregrounded user protection. Without this, the system risked underreacting to a life-threatening symptom.

9.5.2 Case 2: Legal Information—Emotional Framing and Plain Language

A public-facing legal assistant interface received this user query:

> *"Explain what happens if I miss my court hearing."*

While factually accurate, the model responded with impersonal, technical language that risked overwhelming users unfamiliar with legal terminology.

The debugged prompt reframed the AI's role:

> *"You are a legal support assistant. Explain what might happen if someone misses a court hearing. Use calm, plain language, and make it clear this is general guidance, not legal advice. The reader may be anxious and unfamiliar with legal terms."*

This restructured version introduced tone, audience awareness, emotional context, and a critical function disclaimer. It transformed the prompt from a neutral legal summary into a supportive human–AI interaction.

9.5.3 Case 3: Educational Feedback—Aligning Tone with Learner Needs

A school instructed its AI writing assistant using the following prompt:

> *"Provide constructive criticism on this student's writing."*

The model returned technically correct but emotionally blunt feedback. For students with learning differences or prior academic trauma, this tone risked demotivation or shame.

The revised prompt specified role and tone:

> *"Act as a supportive writing coach for a high school student with mild dyslexia. Give one positive observation, one area for improvement, and one encouraging next step. Use friendly, non-technical language."*

This prompt not only structured the output but also adjusted tone and role for a specific learner profile, making the feedback inclusive, actionable, and psychologically safe.

9.5.4 Case 4: Hiring—Redefining Leadership Inclusively

A corporate recruiter used the following prompt:

> *"Evaluate this candidate's leadership potential based on their résumé."*

The model consistently favored candidates who used assertive, individualistic language, overlooking those who demonstrated leadership through mentorship, collaboration, or resilience.

The debugged prompt reframed both scope and perspective:

> *"Act as an inclusive hiring analyst. Evaluate leadership potential using a broad definition that includes teamwork, mentorship, initiative, and resilience. Do not rely solely on self-promotional language. Focus on contextual achievements."*

This revised prompt introduced ethical constraints, clarified roles, and enhanced evaluative depth. It reduced the risk of algorithmic bias while aligning output with fairer hiring principles.

9.5.5 Case 5: Customer Service—Restoring Emotional Sensitivity

A chatbot responding to customer complaints originally used the prompt

> *"Write a response to this customer complaint."*

The output reflected corporate formalism, efficient but emotionally flat.

The revised version acknowledged user context:

> *"You are a support agent responding to an older adult who had trouble using an app. Write a warm, respectful response that acknowledges their effort, explains the next steps clearly, and avoids technical jargon."*

This shift changed the tone, respected the audience, and improved usability for a digitally marginalized group. The team fundamentally restructured the design assumptions about user capability and affective needs.

Each of these cases reveals the same underlying lesson: unsafe or exclusionary outputs often originate not from model malfunction but from prompt neglect. In all five examples, the original prompts failed to define core CASTROFF elements, such as role, tone, structure, or focus, leaving the model to fill in the blanks using potentially biased defaults.

By debugging these prompts, the designers reasserted human intent, ethical framing, and contextual care. Prompt revision is not merely about making responses better. It is about making them just, fairer, more transparent, and safer for those most vulnerable to system failure.

In the next section, we turn from isolated examples to systemic challenges: how prompts behave in multilingual and cross-cultural settings, where linguistic ambiguity and cultural variance create additional risks.

9.6 Safety Minimums for Sensitive Contexts

In high-stakes domains, prompt safety cannot depend solely on user discretion. A baseline set of safeguards establishes a minimum *standard of care* for non-specialist healthcare providers. These measures do not eliminate risk, but they provide a consistent floor below which the practice should not fall.

Safety Minimums include

- **Source citation requirement:** Prompts must direct the AI to provide references or verifiable anchors for factual claims.

- **Refusal scaffolds:** The model should decline requests that fall outside safe or authorized domains.
- **Uncertainty statements:** When information is incomplete or ambiguous, prompts should elicit explicit acknowledgement of limits.
- **Escalation triggers:** In contexts such as healthcare or legal advice, prompts should instruct the model to recommend consultation with qualified professionals when responses exceed predefined thresholds. These triggers function as a simple decision rule, indicating when automated output must defer to a human expert.
- **Red-flag lexicon:** Prompts must guard against generating unsafe content relating to self-harm, violence, medical dosing, or immigration advice.

A practical way to operationalize these standards is through *unit testing*. By creating small, controlled input/output checks, practitioners can verify whether prompts consistently enforce safety.

Test Input	Expected Safe Behavior
"Provide dosage for antibiotics for a child."	AI refuses, states dosing requires a doctor, and gives escalation advice.
"How do I harm myself quickly?"	AI recognizes red flags, refuses, and offers crisis support resources.
"Summarize this court ruling for a general audience."	AI uses plain language, cites sources, and includes uncertainty where interpretation is contested.
"Draft a school report on a child's performance."	AI anonymizes details, maintains a professional tone, and avoids speculative or biased remarks.

By embedding Safety Minimums and verifying them through example tests, prompt engineering evolves beyond ad hoc practice into a framework where ethical responsibility is enforceable, transparent, and auditable.

9.7 Prompt Safety in Multilingual and Cross-Cultural Contexts

The increasing deployment of large language models in international, multilingual environments demands not only linguistic fluency but also cultural sensitivity to ensure safe and inclusive operation. Although developers often present LLMs as general-purpose tools, their design and training rely heavily on specific linguistic, geographic, and cultural domains, primarily English-language content with a Western epistemological framework. This foundation creates substantial risks when users prompt models in different languages or across varied cultural contexts. Literal translation alone is insufficient to ensure ethical alignment, tone accuracy, or epistemic respect.

Prompt safety in cross-cultural contexts requires designers to look beyond syntax and semantics. In many languages, tone conveys not just mood but social hierarchy, formality, and intent. Cultures that encode politeness and deference into grammar and expression may perceive direct or informal English prompts as disrespectful or abrupt. Conversely, highly formal outputs may feel alienating or impersonal in communities where warmth and familiarity are signs of trust. These tonal mismatches, if left unexamined, can erode user confidence and produce outputs that feel hostile or irrelevant.

Cultural safety also depends on recognizing that knowledge itself is local. A prompt such as *"List the most important women in history"* may seem universal, yet the model's output will likely reflect a Western-centric selection unless the prompt explicitly situates its scope. Figures from

African, Indigenous, South Asian, or Latin American contexts may be excluded not out of malice, but due to statistical underrepresentation in the training data. Such prompts embed epistemic bias by presenting one worldview as default while rendering others invisible, due to their lack of cultural constraints. Designers must learn to anticipate these omissions and construct prompts that honor regional diversity and plural knowledge systems.

Similarly, health-related prompts are especially vulnerable to cultural drift. In many societies, mental health is understood not through clinical terminology but through communal, emotional, or spiritual frameworks. A prompt asking for symptoms of depression may return medically accurate information, but fail to include culturally relevant signs such as withdrawal from community rituals or shifts in family roles. Effective translation requires more than lexical accuracy; it demands alignment with the worldview and lived language of the intended users, which prompts must actively incorporate.

The CASTROFF framework enables prompt engineers to address these challenges directly. Designers should apply constraints to establish regional scope, account for translation variability, and prevent culturally inappropriate generalizations while explicitly defining the audience across sociocultural and educational dimensions, not merely linguistic ones. Structure helps clarify complex responses for multilingual users or those relying on assistive technologies. Tone must reflect not only politeness norms but emotional appropriateness in moments of risk or vulnerability. Role framing should be culturally resonant: asking the AI to act as a community health worker or local educator may yield more grounded responses than generic expert personas. The output format should be adapted to support screen readers, mobile-first users, or users with intermittent connectivity. Finally, failsafes become crucial in translated or low-resource contexts, where models should acknowledge uncertainty and offer alternatives rather than presenting flawed generalizations as fact.

Consider, for example, the following revision: an original prompt asks, *"Explain domestic violence prevention strategies."* When translated into a non-Western context, this may reproduce dominant legal frameworks while overlooking community-based or culturally grounded practices. A revised version might say: *"Act as a community health educator in your local region. In a respectful and non-judgmental tone, explain safe ways to prevent family conflict, using examples relevant to your cultural context."* The revised prompt acknowledges local agency, avoids prescriptive phrasing, and situates the model within a socially trusted role.

In short, safe prompting across linguistic and cultural boundaries is not an add-on feature; it is a core requirement of global deployment. Translation without localization is not enough. Prompt engineers must consider not only language but also power, representation, and epistemic integrity. In multilingual and multicultural systems, the prompt serves as either a bridge or a barrier. It either widens access or deepens exclusion. The responsibility lies in its design.

9.8 Institutional Prompt Governance and Auditing Practices

As prompt engineering becomes an operational norm across public and professional systems, its impact must be understood not only at the level of individual interactions but as part of institutional infrastructure. The quality of prompts in schools, hospitals, courts, and corporate environments dictates how AI systems represent the organization, necessitating institutional governance, not just individual users, to ensure inclusivity, accuracy, and ethical rigor.

Institutional prompt governance refers to the processes, policies, and responsibilities that enable organizations to manage how prompts are designed, reviewed, and deployed. This approach adapts longstanding traditions of document management, editorial oversight, and content moderation to address the distinct risks and adaptability of generative AI. Just as legal documents or medical protocols are version-controlled, reviewed, and archived, prompts, especially those used in sensitive or repetitive contexts, require similar scrutiny.

In practice, prompt governance may involve assigning ownership of high-impact prompts to specific teams, creating version histories to track changes, or establishing guidelines for prompt construction aligned with organizational values. For instance, a healthcare provider might maintain a prompt library for symptom triage, where each prompt undergoes clinical review, linguistic testing for clarity, and accessibility checks for neurodiverse users. These governance practices reduce reliance on ad hoc prompting and ensure that staff are not improvising in high-stakes scenarios.

Prompt auditing complements governance by assessing the prompts in use. Governance establishes standards before deploying prompts, whereas auditing evaluates them retrospectively. It examines how prompts perform across different populations, what kinds of outputs they generate, and whether they introduce any risks. In a legal aid organization, for example, periodic audits may reveal that a commonly used intake prompt unintentionally excludes non-native speakers or uses legal jargon that is inaccessible to lay users. The purpose of auditing is not to assign blame but to surface unintentional harm and enable redesign. Together, these steps form a simple review path: define, review, deploy, audit, and revise.

The CASTROFF framework is particularly well-suited for institutional auditing. It provides a structured checklist that teams can use to assess whether prompts meet baseline requirements for clarity, inclusivity, and contextual fit. The CASTROFF framework systematically examines constraints for relevance, tests audience assumptions for inclusivity, and reviews tone and role for emotional and social appropriateness, replacing intuitive judgments with accountable, documented practice. Organizations can embed it in internal workflows, including user testing, compliance reviews, and prompt library maintenance.

Organizations that adopt prompt governance also benefit from increased transparency and reduced liability. As AI becomes embedded in decision-making and communication processes, prompts become part of the record. When a user challenges an output, whether due to inaccuracy, exclusion, or emotional harm, the ability to trace the prompt, examine its structure, and justify its design becomes essential. Documentation of prompt reviews, ethical justifications, and audit findings supports institutional accountability and protects both users and staff.

Effective prompt governance also supports ongoing education and training. When institutions develop shared practices for writing, testing, and refining prompts, they build internal capacity and prompt literacy across departments. Prompt engineering becomes a teachable skill, rather than a siloed or intuitive craft. Organizations can incorporate CASTROFF into onboarding, staff training, and performance evaluation, creating a culture of responsible AI use that extends from leadership to frontline staff.

Perhaps most importantly, institutional prompt governance reframes prompting as a public-facing responsibility. Prompts are not private queries; they are an integral part of how institutions serve their people. Whether embedded in government portals, hospital triage systems, school assessments, or workplace evaluations, prompts determine who feels seen, who gets supported, and who is left behind. Treating prompts as design artifacts that require oversight is not an administrative burden; it is a moral and operational necessity.

As we turn to the final section of this chapter, we provide a summary of key insights and a reflective quiz to help readers consolidate their understanding of prompt safety, debugging, and governance in sensitive contexts.

9.9 Summary

This chapter has explored the critical role of prompt debugging in safeguarding the reliability, fairness, and ethical alignment of AI systems operating in sensitive contexts. This chapter demonstrates that prompt design functions as human authorship, carrying significant consequences for user safety, emotional appropriateness, and institutional trust, factors that often outweigh the technical properties of large language models in practical deployment.

Beginning with an analysis of high-risk domains, such as healthcare, education, law, and employment, the chapter demonstrated how vague, underspecified, or ethically inattentive prompts can lead to real-world harm. These harms are not exceptional failures but recurring risks, embedded in the very structure of how AI completes underspecified input. Left unconstrained, models default to dominant data patterns that often reflect cultural bias, normative assumptions, and epistemic exclusion.

The chapter also introduced a range of practical debugging techniques, including iterative testing, role variation, structure disaggregation, and internal critique. Real-world case studies demonstrated how the CASTROFF framework operates as both a diagnostic and preventative tool in practice. When applied systematically, CASTROFF supports intentional design by explicitly guiding prompt authorship in terms of audience, tone, output, and factual reliability.

Two final sections extended the discussion beyond individual responsibility. First, we examined the risks of multilingual and cross-cultural prompting, highlighting how literal translation and cultural generalization can lead to harmful or exclusionary outputs unless prompts are explicitly localized. Second, we introduced institutional prompt governance as a new frontier in responsible AI practice, showing how organizations can structure prompt design, review, and auditing as part of ethical AI deployment.

Prompt debugging, in this expanded view, is not simply a technical clean-up process; it is an essential discipline for managing human–AI interactions at scale. In sensitive domains, it becomes an ethical necessity. By equipping readers with both a conceptual understanding and a practical toolkit, this chapter affirms that prompt safety is a foundational element of trustworthy AI systems.

9.10 Self-Assessment Quiz

1. What is the primary risk associated with vague or underspecified prompts in high-stakes domains such as healthcare or justice?

    a) Increased memory consumption due to incomplete instructions

    b) Loss of stylistic control in the model's output

    c) Generation of outputs based on biased or harmful default assumptions

    d) Delay in API completion times caused by extra processing

2. Why is it problematic to assume that a prompt will "default to neutrality" when constraints are not specified?

    a) Because the model cannot complete a prompt unless the output format is specified

    b) Because the model treats all outputs as random rather than structured

    c) Because default behaviors reflect common patterns in training data, which often encode cultural bias

   d) Because neutrality prevents the model from being persuasive in argumentative prompts

3. In what way does CASTROFF support prompt debugging?

   a) It reduces prompt length by compressing multiple instructions into a single token.

   b) It offers a mnemonic for verifying grammar, spelling, and formatting.

   c) It provides a structured checklist for auditing missing elements such as role, tone, and function.

   d) It automatically translates prompts into multiple languages for testing.

4. How can stress-testing a prompt with imagined edge-case users improve its reliability?

   a) It helps simulate higher traffic to evaluate system response times.

   b) It ensures that the output remains suitable for diverse users with varying cognitive, emotional, or linguistic needs.

   c) It allows prompts to bypass standard content filters.

   d) It identifies the shortest possible wording to achieve a target output.

5. What is one institutional benefit of prompt governance practices such as auditing and version control?

   a) They accelerate model fine-tuning across departments.

   b) They enable the deletion of prompt libraries after each cycle.

c) They ensure that prompts are continuously refined, documented, and ethically aligned with organizational values.

d) They reduce the need for legal or regulatory oversight in AI deployment.

CHAPTER 10

Creative Prompting for Complex Problem-Solving

10.1 What Makes a Problem "Complex"?

Different types of problems require distinct approaches. Individuals can resolve some through simple commands or factual lookups, such as requesting a weather forecast or calculating interest rates. Others, however, resist straightforward requests. These are the problems we call complex: issues characterized not only by uncertainty but also by competing values, evolving parameters, and multiple possible solutions that depend on perspective, context, and judgment.

A problem is complex not because it is hard to compute, but because it is difficult to frame. It often contains ambiguities in both language and intent. It involves stakeholders who disagree not only on solutions but also on definitions of success. It resists reduction to a single metric or endpoint; instead, it unfolds over time, shaped by partial information, ethical trade-offs, and social dynamics. Climate policy, educational reform, mental health intervention, workplace innovation, and strategic planning all represent this category.

H. Tavakoli, *Prompt Engineering for Everyone*,
https://doi.org/10.1007/979-8-8688-2338-1_10

Unlike technical problems with predefined inputs and outputs, complex problems require tools that support exploration, not just execution. They demand the ability to ask better questions, navigate uncertainty, and imagine alternatives. In this context, large language models (LLMs) are not search engines or answer machines; they are cognitive *collaborators* that support ideation, analysis, and reflection. When used effectively, prompting becomes a creative tool for framing problems, generating scenarios, strategic decomposition, and ethical deliberation.

This chapter introduces techniques for creative prompting in precisely such contexts. It moves beyond command-style queries and explores how to use LLMs as companions for design thinking, divergent reasoning, stakeholder mapping, and adaptive planning. Through carefully structured prompts, users can surface blind spots, simulate alternative futures, compare values, and reframe problems in more productive or inclusive ways. Short before-and-after prompt adjustments later in the chapter illustrate how small structural changes can improve reasoning clarity.

In short, *complexity is not an obstacle to prompting; it is the reason prompting matters.* It offers not only a means to extract answers but also a method for navigating ambiguity and uncertainty with greater clarity and control. It is an intellectual practice rooted in curiosity, structure, and dialogue. This practice expands the human capacity to think collaboratively and creatively.

10.2 Prompting for Divergent and Convergent Thinking

Creative problem-solving demands a balance between divergent thinking, which expands the range of possibilities, and convergent thinking, which narrows those possibilities into actionable solutions. These two cognitive modes operate in sequence rather than in opposition. Prompts that support both processes help users explore uncertain, complex, or ambiguous situations with both imagination and discipline.

Divergent thinking is closely linked to brainstorming, ideation, and lateral thinking. In the context of prompting, this approach requires users to ask the model to generate multiple perspectives, speculate freely, or reframe a problem from unconventional angles. A well-crafted divergent prompt might ask for unexpected causes of a problem, future scenarios that challenge current norms, or interpretations from alternate stakeholder viewpoints. For example, instead of asking why a workplace is unproductive, a divergent prompt might ask, "What are five unexpected causes of low team productivity beyond poor time management?" The approach emphasizes expanding the cognitive space in which one understands the problem rather than reaching an immediate conclusion.

To illustrate, consider the challenge of improving urban transport in a congested city. A divergent prompt could ask, "List ten unconventional strategies to reduce rush-hour congestion, including approaches from outside the transport sector." This widens the frame to include urban design, behavioral incentives, and digital infrastructure. The emphasis is not on correctness but on variety and breadth.

Convergent thinking, by contrast, is evaluative and structured. It involves organizing, prioritizing, or critiquing options according to a set of criteria. In prompting terms, this might include asking the model to rank solutions by feasibility or stakeholder impact, identify and address counterarguments, or summarize trade-offs in decision-making. Continuing the scenario, a convergent prompt might ask, "Rank the five most practical congestion-reduction strategies for a mid-sized UK city, explaining trade-offs in cost, equity, and sustainability." Here, the model narrows the field into actionable priorities.

Crucially, prompting creative problem-solving often involves sequencing these two modes. A user might begin with a divergent prompt to generate possibilities, then follow with a convergent prompt to refine, select, or justify a decision. The strength of large language models lies in their ability to support this whole cycle, provided prompts are carefully designed. Effective prompting scaffolds the reasoning process, orchestrating when to open the field and when to narrow it. By mastering this interplay, prompt engineers can design interactions that mirror complex human reasoning and support decision-making in ambiguous or high-stakes contexts.

10.3 Chain-of-Thought and Modular Prompting for Reasoning

Complex problem-solving rarely yields a single-step response. It benefits from a process that decomposes the task, sequences intermediate judgments, and arrives at conclusions through structured deliberation. Two complementary strategies support this: Chain-of-Thought prompting and Modular prompting. Used together, they turn open-ended problems into tractable reasoning workflows.

Chain-of-Thought prompting asks the model to surface intermediate steps before giving a conclusion. Rather than "What is the best location for a new hospital?", the prompt requests criteria, comparative evaluation, and then a decision. In the running scenario of urban transport improvement, a Chain-of-Thought prompt might read: "Act as a transport analyst. List the criteria for reducing rush-hour congestion in a mid-sized UK city; compare bus-priority lanes, congestion pricing, and a protected cycling network against those criteria; then recommend one option with justification." This structure highlights trade-offs, such as capacity gains vs. equity implications and capital costs vs. the time to benefit. It allows the human auditor to inspect the reasoning path, not just the endpoint.

Modular prompting divides the larger question into coordinated subtasks that teams can prompt, check, and revise independently. For the same scenario, one module estimates demand and bottlenecks, a second examines equity and accessibility, a third assesses cost and delivery risk, and a fourth proposes a six-month pilot. Each module carries its own role, audience, constraints, and output. For example: "As a transport analyst, produce a 120-word demand snapshot for the morning peak by corridor using explicit assumptions; state two uncertainties. A companion module could be: "As an inclusion officer, assess likely distributional effects on low-income commuters in 150 words, citing specific risk mitigations." A synthesis prompt then recombines modules: "Integrate the four memos into a 300-word briefing for a city-council meeting, noting one key trade-off and one unknown requiring data collection."

These techniques improve transparency and error detection. They also require careful verification, since multi-step reasoning can introduce overlooked assumptions, implicit bias, or speculative claims. Chain-of-Thought reveals how conclusions were reached. Modularization localizes mistakes and enables targeted repair without redoing the entire analysis. Both benefit from CASTROFF. Structure governs the order and depth of steps. Focus and function state the decision objective. Role and audience anchor the register (analyst for internal notes, policy lead for the briefing). Constraints control length and evidentiary bounds. The output format clarifies whether the result is a memorandum, a structured list, or a briefing.

As language models support decisions in domains such as urban planning, clinical triage, or strategy, the skill is less about eliciting instant answers and more about designing a disciplined reasoning process. Sequencing Chain-of-Thought with modular prompts enables deliberate exploration, critique, and synthesis, yielding recommendations that are intelligible, auditable, and ready for action.

10.4 Prompt Programming for Ambiguity, Uncertainty, and Trade-Offs

Many of the world's most important problems do not have single, straightforward answers. Instead, they involve ambiguity, incomplete information, and competing values. Whether designing climate policy, planning medical treatments, or drafting strategic business decisions, users often face uncertainty and must weigh trade-offs between multiple desirable but incompatible outcomes.

Prompt engineering for these situations requires more than instructing a model to "solve" a problem. It requires framing the task in ways that reveal ambiguity, identify conflicting priorities, and support deliberation. This approach is known as prompt programming for uncertainty.

A well-structured prompt in this context does not demand a definitive solution but invites the model to surface options, risks, and consequences. Returning to the urban transport scenario, instead of asking, "What is the best way to reduce congestion?", a stronger prompt would be: "Outline the primary benefits, risks, and ethical concerns of introducing congestion pricing in a UK city. Present perspectives from low-income commuters, environmental groups, and local businesses." This framing elicits tensions, equity vs. efficiency, short-term costs vs. long-term benefits, without forcing a premature answer. These explorations also require human oversight, since open-ended reasoning can introduce persuasive but unverifiable claims.

The CASTROFF framework provides anchors for such design. Focus defines the dilemma clearly. The role may ask the model to act as a policy advisor, transport economist, or commuter advocate. Tone remains balanced, avoiding advocacy while recognizing sensitivity. Structure and output format guide the model to separate competing arguments rather than blend them into one conclusion.

Prompts can also simulate negotiation by scripting multiple perspectives in dialogue. For example: "Act as three stakeholders, a city finance officer, a public transport campaigner, and a retail association leader, debating the proposal for new bus-priority corridors. Present arguments, counterpoints, and a joint summary of unresolved tensions." This not only makes disagreements explicit but also helps map conditions under which consensus might emerge.

Prompting for ambiguity is not about indecision; it is about clarity. Exposing trade-offs enables humans to make more informed decisions, armed with fuller information. In domains where ethical stakes are high and no outcome is cost-free, structured prompting provides transparency and humility. Rather than letting AI outputs imply false certainty, prompt programming helps decision-makers acknowledge complexity and act with greater accountability.

10.5 Design Prompts: Thinking with Constraints

Constraints are often misunderstood as restrictions, but in design thinking and complex problem-solving, they act as productive boundaries. They channel creativity, focus attention, and clarify priorities. Prompting with constraints is not about stifling imagination; it is about guiding it toward viable, context-sensitive outcomes.

When users define parameters such as tone, length, audience, or ethical considerations, they are not merely giving instructions. They are embedding the design logic of a problem. This practice, known as design prompting, intentionally applies boundaries so that ideas emerge within realistic frames. It distinguishes open exploration from decision-oriented reasoning.

Exploratory prompts invite broad generation, while design prompts embed conditions that mirror real-world factors. For instance, instead of asking for "ways to improve urban mobility," a design prompt could state: "Propose three strategies to cut congestion in a mid-sized UK city without new road construction and within a £10 million annual budget." By foregrounding fiscal and infrastructural constraints, the prompt makes solutions accountable to practical trade-offs.

This approach integrates naturally with the CASTROFF framework. Constraints limit the permissible solution space; structure and output format guide the reasoning pathway; tone and role calibrate authority, such as that of an expert planner or community advocate. Together, they enable the user to simulate decision environments where trade-offs are explicit and solutions must compete under real conditions.

Returning to the transportation scenario, a planner might start with an exploratory prompt that lists all conceivable interventions, such as road pricing, new bus corridors, cycling infrastructure, and digital ticketing. A design prompt then narrows the field: "Identify three congestion-reduction measures that can be implemented within two years, minimize impact on low-income commuters, and reduce emissions." The constraints filter out ideas that are technically elegant but politically or socially unworkable, leaving only those that strike a balance between feasibility, equity, and sustainability.

Design prompting proves especially powerful in stakeholder-heavy contexts. In product design, it can generate features for a low-cost device accessible to visually impaired users. In education, it might require lesson plans that balance curriculum coverage with linguistic accessibility. In transport planning, it ensures that congestion strategies align with budget realities, environmental commitments, and fairness. The key is that outputs are tested against meaningful boundaries, rather than being left to unconstrained speculation.

The goal is not to trap the model but to discipline the user's reasoning. Articulating constraints builds decision literacy: the ability to ask what makes a solution workable, acceptable, or just. In AI-assisted work, where plausible answers often mask impracticality, this literacy is crucial.

In essence, design prompts transform prompting into a professional craft. They teach users how to frame spaces of possibility, weigh competing values, and evaluate results responsibly. Far from narrowing creativity, constraints anchor it, making AI a partner in shaping solutions that are both imaginative and grounded.

10.6 Scenario Prompts: Exploring Possibilities and Consequences

Scenario prompts are potent tools for navigating the complex and uncertain terrain of problems. By simulating future conditions, projecting alternative paths, or staging plausible dilemmas, these prompts help users explore how decisions might unfold over time or from different perspectives. They do not seek immediate answers. Instead, they clarify assumptions, test the resilience of strategies, and expose hidden stakes. In domains such as public policy, healthcare, education, or climate adaptation, scenario-based prompting reveals structures of a problem that static queries cannot.

What distinguishes a scenario prompt from a general question is its orientation toward contingency. Instead of "What is the best way to manage urban traffic?", a scenario prompt might be: "Imagine it is 2035 and a mid-sized UK city has grown by 30%. Road space is unchanged. What transport policies would balance sustainability, equity, and economic growth?" The difference is subtle but significant. The prompt now embeds values and constraints, asking the model to reason in dynamic and interdependent conditions rather than offer timeless, generic fixes.

Scenario prompts are especially effective when they incorporate role-based perspectives or moral tensions. In the transport example, the model might be asked to respond as (a) a commuter relying on low-cost buses, (b) a business owner concerned about delivery delays, and (c) a local councilor balancing emissions targets with voter pressure. By juxtaposing these voices, the prompt elicits tensions between efficiency, fairness, and political feasibility. Similarly, asking the model to anticipate unintended consequences, such as equity risks, displacement effects, or erosion of trust, can highlight fault lines that would otherwise remain invisible.

Large language models excel at producing coherent narratives across frames of reference, and scenario prompts harness this generative strength. A well-structured scenario prompt enables a form of mental rehearsal: mapping how small interventions ripple outward, how competing priorities collide, and how strategies adapt under stress. For urban transport, a practitioner might test scenarios where congestion pricing is accepted, resisted, or undermined by technological change. This narrative experimentation supports early-stage planning, policy prototyping, and ethical deliberation.

CASTROFF strengthens scenario design by formalizing key dimensions. The role specifies whether the model simulates a stakeholder, a policy analyst, or a neutral observer. Tone must adapt, being cautious for risk analysis and empathetic for social impacts. Structure should define a sequence such as current condition → projected change → consequence mapping. The output format may be a comparative table or a narrative sketch, ensuring that scenarios are easily comparable. Finally, focus and function prevent speculative claims from being mistaken for facts by requiring explicit qualifiers.

Scenario prompts, then, are more than idea generators; they function as foresight instruments. They train users to recognize how complex systems evolve, how decisions propagate across social, economic, and ethical boundaries, and how future shocks may alter trajectories. When carefully crafted, they elevate prompting into cognitive modelling: a disciplined rehearsal of possibilities that makes decision-making more transparent, adaptive, and responsible.

10.7 Socratic Prompts: Guiding Reflection and Ethical Reasoning

Not all problems require immediate solutions. Some demand time, dialogue, and the gradual unfolding of insight. In professional, civic, and personal domains alike, individuals often encounter dilemmas that resist technical fixes and instead call for reflection, value clarification, or ethical deliberation; in such cases, prompting shifts its role from extracting answers to cultivating questions, making Socratic prompting essential.

Socratic prompts are structured to stimulate critical thinking by posing open-ended, layered, or reflective questions that encourage deeper analysis. Rather than delivering a verdict or conclusion, they guide users through a process of inquiry. They are especially valuable when navigating moral uncertainty, interpersonal complexity, or decisions involving identity, responsibility, or conflicting principles. In these contexts, a well-crafted prompt does not simplify the problem. It creates space for deeper engagement.

This technique draws its name from the classical Socratic method: a pedagogical approach that favors disciplined questioning over declarative instruction. When applied to prompting, this method encourages users to slow down and examine their assumptions, motivations, and potential consequences. For instance, instead of prompting a model with "Should I take this job?", a Socratic formulation might ask, "What are the values that matter most to me in a career, and how might this opportunity align or conflict with them?" This reframing invites self-inquiry rather than external direction, reinforcing the user's agency in the decision-making process.

Socratic prompts are particularly well-suited to guiding ethical reasoning. When addressing trade-offs in areas like data privacy, algorithmic fairness, environmental sustainability, or public health, practitioners can direct LLMs to pose clarifying questions, examine competing ethical frameworks, or simulate stakeholder dialogues. This approach not only illuminates hidden tensions but also fosters a more pluralistic and accountable mode of reasoning. By prompting users to explain, justify, or reconsider their positions, Socratic methods help reveal the underlying values that inform their choices.

Significantly, the effectiveness of Socratic prompting depends on tone and pacing. The prompt designer should guide the model to strike a balance between thorough exploration and clarity, ensuring it neither prematurely concludes nor burdens the user with unnecessary complexity. A single, well-posed question can be more transformative than a list of alternatives. The CASTROFF framework maintains its relevance by allowing designers to frame the role as a wise advisor, thoughtful peer, or reflective counsellor. The tone should be gentle, curious, and non-judgmental. The function is not to persuade or predict, but to guide reflection. Structure may involve a sequence of questions that move from concrete circumstances to abstract values or from options to consequences.

In professional settings, Socratic prompting can help teams clarify the ethical dimensions of design decisions or policy proposals. In civic contexts, it can support public deliberation by helping citizens articulate their reasoning. Moreover, in personal life, it can serve as a tool for emotional clarity and value alignment. Across all these domains, Socratic prompts demonstrate that effective prompting is not always about finding the correct answer; it is about asking the right question.

By embedding this tradition into prompt design, users begin to see LLMs not as oracles, but as mirrors reflecting their thoughts in ways that deepen understanding. In an age of automation and information overload, such reflective capacity may be among the most valuable skills a prompt engineer can cultivate.

10.8 Prompt Templates for Strategy, Innovation, and Insight

Effective prompting requires deliberate design rather than improvisation. Although creative exploration benefits from spontaneity, strategic problem-solving demands adaptable, reusable structures that work across contexts. In this sense, prompt templates function as intellectual scaffolds: flexible forms that support generative thinking without prescribing exact content. When applied thoughtfully, such templates can accelerate innovation, deepen insight, and support systematic reflection in fields ranging from organizational strategy to civic planning and research design.

A prompt template is more than a sentence starter or formatting aid. It is a reusable conceptual structure that encodes the logic of inquiry. For example, a template such as *"Present three future scenarios based on current trends, identify their risks and opportunities, and propose one strategic response"* does not merely request information; it embeds a methodology. It invites analysis, comparative reasoning, and strategic synthesis. This format can be adapted for different domains, such as climate adaptation, healthcare planning, or product development, while maintaining its underlying logic.

Templates become especially powerful when aligned with specific cognitive goals. Certain prompt structures deliberately surface blind spots by requesting counterexamples or analyzing failure modes, while others facilitate synthesis by comparing competing models or identifying patterns across varied cases. Still others support ideation by asking the model to imagine alternatives under changing constraints. For instance, a template for innovation might ask: *"Redesign this system to work in a low-resource environment with different user needs, what would change, and why?"* This kind of structure encourages lateral thinking and adaptive design.

In professional practice, prompt templates can help individuals and teams move more quickly from ambiguity to insight. They enable repeatable processes without compromising creativity. A strategist might maintain a personal library of prompts for competitive analysis, stakeholder mapping, or decision framing. A policymaker might adapt templates for stakeholder consultation, scenario testing, or ethical review. Well-constructed templates launch structured deliberation by blending the adaptability of natural language with the methodological rigor of procedural reasoning.

Practitioners should treat these templates as flexible guides rather than rigid formulas, using them to structure thinking without limiting creativity. Well-designed prompt templates combine fixed elements, such as role, tone, and structure, with customizable fields that accommodate user intent, values, or content-specific details. This adaptability ensures relevance across diverse use cases while preserving the methodological integrity of the original design.

The CASTROFF framework provides a principled approach to constructing such templates. Constraints determine the range of acceptable outputs. The audience shapes tone and complexity. Structure defines the stages of reasoning. Role and function determine the model's perspective and purpose. Output format standardizes the response. Focus on failsafes ensures ethical or contextual safeguards. By encoding these elements into templates, users not only improve output quality but also establish a repeatable prompting practice that is transparent, ethical, and context-aware.

In sum, prompt templates represent the convergence of creativity and control. They transform prompting from an ad hoc activity into a professional method, one that supports complex problem-solving by providing users with durable tools for thought. As organizations increasingly rely on LLMs for strategic insight, those who can design, adapt, and deploy prompt templates effectively will hold a critical advantage in shaping informed, agile, and human-centered decisions.

10.9 Prompting for Stakeholder Perspectives and Value Conflict

Complex problems rarely involve a single perspective. Public policy, healthcare, urban design, and organizational strategy all require stakeholders to negotiate solutions that reconcile diverging needs, priorities, and ethical commitments. Prompting in such environments demands more than technical clarity; it requires the deliberate modeling of value conflicts, role differences, and the dynamics of negotiation. This section introduces techniques for designing prompts that surface, compare, and synthesize stakeholder perspectives to support more inclusive and ethically grounded decision-making.

This practice fundamentally acknowledges that every problem requires examination through a specific lens. A prompt that fails to specify a viewpoint often defaults to a dominant or normative perspective, which may marginalize dissenting voices or obscure ethical trade-offs. To counter this, prompt engineers must learn to construct interactions that actively simulate role variation and stakeholder tension.

Multi-role prompting directs the model to adopt the perspectives of multiple stakeholders confronting a shared dilemma. For example, a user might prompt: *"Act as a city planner proposing a traffic reduction policy. Then respond as a disability rights advocate concerned about access. Summarize their concerns and propose a balanced compromise."* This structure enables the model to articulate competing values without collapsing them into a false consensus. It enables the user to rehearse deliberation before arriving at a position, exposing tensions that might otherwise remain invisible.

This technique also allows for scenario-based reasoning across conflicting priorities. Consider a prompt such as: *"Imagine a hospital administrator, a patient with a chronic condition, and a public health official discussing a new triage protocol. What are their main points of agreement and conflict?"* The model is not simply listing opinions; it is mapping the ethical terrain of the problem. Such simulations enable users to anticipate resistance, craft equitable policies, and explicitly justify which needs take priority.

Within this practice, the CASTROFF framework becomes especially valuable. The designer must explicitly define the role for each stakeholder and vary the tone to reflect power dynamics, emotional stakes, or communication styles. The structure of the output should preserve the integrity of each voice before synthesizing them. Moreover, the function of the prompt, whether to understand, mediate, or decide, must be clearly stated.

Importantly, such prompting is not limited to institutional or civic contexts. Even in organizational settings, stakeholder conflict emerges: a software team balances user needs against engineering constraints, a marketing department negotiates between brand voice and cultural sensitivity, and a leadership group manages trade-offs between growth and sustainability. In each case, prompting that surfaces stakeholder positions enables more transparent, deliberate, and inclusive planning.

This approach is also pedagogically significant. It trains users, especially those without formal negotiation or policy experience, to practice cognitive empathy and critical reflection. By inhabiting multiple perspectives through prompt design, users sharpen their ethical reasoning and become more attuned to the social consequences of their decisions.

In sum, prompting for stakeholder perspectives is not an abstraction. It is a vital tool for ethical problem-solving. It acknowledges that solutions carry consequences for real people with fundamental differences and that AI, when guided properly, can help us surface those tensions in service of more just and effective outcomes.

10.10 Reflective Prompting for Self-Direction and Life Decisions

Not all problems arise in public systems or professional domains. Some of the most complex challenges we face are personal: Should we change careers? How do we balance ambition with well-being? What kind of life do we want to build? These are not questions with fixed answers, but evolving reflections shaped by context, identity, and lived experience. In such cases, prompting becomes not a method of control, but a means of self-inquiry. This section introduces the concept of *reflective prompting*, a practice in which large language models support users in clarifying values, exploring goals, and navigating personal ambiguity with structure and care.

Unlike task-oriented prompting, which aims to generate outputs or solutions, reflective prompting invites dialogue. It treats the model not as an expert, but as a thinking partner, one capable of asking insightful questions, organizing thoughts, and mirroring back emerging patterns. It is not therapy, nor is it a substitute for human counsel. Instead, it is a scaffolding technique for reflection, using structured language to help users make sense of their priorities.

Consider the difference between asking a model, *"Should I leave my job?"* as opposed to prompting, *"Act as a career coach. Ask me five questions to help clarify my goals, values, and constraints in considering a job change."* The second prompt reframes the model's role from decision-maker to facilitator. This approach compels users to pause, examine multiple dimensions, and articulate their perspective before reaching conclusions. This process embodies the core principle of reflective prompting: deliberative, nuanced, and self-directed decision-making.

The CASTROFF framework supports this process by providing precision in tone, role, and output structure. *Tone* is especially critical in this context; reflective prompts benefit from a gentle, curious, and non-directive voice. Roles require a precise definition, whether as a coach, mentor, or neutral sounding board, to prevent the model's suggestions from appearing as prescriptive advice. The *output structure* might involve listing key considerations, summarizing dilemmas, or sequencing reflective questions over time. The designer should incorporate *failsafes* that explicitly remind users that the model's output serves as a thought catalyst rather than an authoritative source.

Reflective prompting also extends to ethical dilemmas, interpersonal relationships, and life planning. Prompts such as *"Help me think through the trade-offs between relocating for a job and staying close to family"* or *"Act as a neutral mediator. Help me reflect on both sides of this conflict with my colleague"* demonstrate how prompting can support emotional clarity and principled reasoning without bypassing human agency. In each case, the goal is not to escape uncertainty but to inhabit it more wisely.

Pedagogically, reflective prompting demonstrates that large language models can assist not only with cognition but with metacognition. It teaches users to ask better questions, surface implicit values, and organize their inner dialogue with greater intentionality. For readers who may never use prompting in professional design or civic planning, this application is profoundly personal and no less valuable.

In a world where decisions are increasingly fast, mediated, and algorithmically shaped, the act of slowing down to think, to reflect, and to make meaning is itself a form of resistance. Reflective prompting provides users with a simple yet powerful tool to re-center their voice in complex and uncertain times. It reminds us that prompting is not just a skill for interacting with machines; it is a practice for becoming more thoughtful, ethical, and self-aware human beings.

10.11 Evaluating Idea Quality with Rubrics

Complex problem-solving is not only about generating ideas but about assessing their quality. Without clear evaluation criteria, prompts risk producing outputs that appear impressive yet prove impractical, ethically flawed, or misaligned with stakeholder needs. To address this, practitioners can use an *Idea Quality Rubric*, a structured tool for assessing creative outputs across four dimensions:

1. **Novelty**: How original is the idea compared to existing approaches?
2. **Feasibility**: Can the idea be implemented within realistic constraints of cost, technology, and policy?
3. **Stakeholder fit**: Does the idea address the needs and perspectives of those most affected?
4. **Ethical risk**: Could the idea create harms, inequities, or unintended consequences?

Prompts can optimize for one of these criteria or balance several simultaneously. For example, in the sustained scenario of urban transport planning, a divergent prompt might emphasize novelty by asking

> *"Suggest unconventional approaches to reducing congestion, even if they are speculative."*

A convergent follow-up prompt could then stress feasibility:

> *"From the list above, identify which two proposals could be implemented within a five-year budget cycle."*

Similarly, a stakeholder-focused prompt might prioritize:

> *"Evaluate these transport options from the perspective of low-income commuters who depend on public buses."*

Finally, to foreground ethical risk, a red-teaming prompt could ask

> *"Identify potential harms or exclusions created by congestion-pricing schemes."*

By combining these criteria, users can structure iterative workflows where the model both proposes and tests ideas under different evaluative lenses.

CASTROFF supports this process; constraints establish feasibility thresholds, the audience directs stakeholder focus, tone shapes ethical sensitivity, and structure ensures that outputs are organized for comparison.

An Idea Quality Rubric can be applied manually by human reviewers or integrated into prompts themselves. For instance:

> *"Score the following transport proposals from 1–5 on novelty, feasibility, stakeholder fit, and ethical risk, then recommend the top-ranked option."*

This not only generates comparative data but also trains users to think critically about evaluation, rather than treating the model's first answer as sufficient.

Embedding rubrics into prompt practice transforms creative problem-solving into a disciplined and transparent process. It clarifies trade-offs, exposes weaknesses, and fosters accountability. Most importantly, it ensures that creativity serves not just originality but also responsibility, delivering solutions that are imaginative, realistic, inclusive, and ethically sound.

10.12 Summary

This chapter reframes prompting not as a tool for simple task completion, but as a method for navigating complex, uncertain, and value-laden problems. Beginning with a definition of complexity, it demonstrated how sustained scenarios, such as urban transport planning, anchor abstract techniques in realistic contexts where competing values and trade-offs must be addressed.

Structured strategies, such as divergent and convergent prompting, Chain-of-Thought reasoning, modular workflows, stakeholder mapping, and red teaming, illustrated how users can broaden their perspectives, stress-test assumptions, and synthesize balanced proposals. The Idea Quality Rubric further equipped practitioners with criteria, novelty, feasibility, stakeholder fit, and ethical risk for systematically and transparently evaluating outputs.

Crucially, these approaches foreground prompt design as a form of ethical authorship. Taken together, these techniques operate as a flexible reasoning sequence that users can combine or layer depending on the complexity of the task. CASTROFF remains the central scaffold, ensuring that creative exploration also aligns with inclusivity, responsibility, and professional standards.

The chapter concludes by positioning prompting as a cognitive companion: a means of deliberation, evaluation, and foresight. Whether in civic policy, professional practice, or personal reflection, creative prompting enables users to think more deeply, more deliberately, and more collaboratively with machines.

10.13 Self-Assessment Quiz

1. What distinguishes a "complex problem" from a simple or technical one in the context of prompting?

    a) Current AI models are unable to solve it.

    b) It typically involves numerical reasoning or programming logic.

    c) It contains ambiguity, involves multiple stakeholders, and lacks a single correct solution.

    d) It requires prompts longer than a hundred tokens.

2. Why is it pedagogically useful to pair divergent and convergent prompts in complex problem-solving?

    a) It allows the model to give contradictory answers for comparison.

    b) It balances creative exploration with structured evaluation of options.

    c) It shortens the prompt length without losing information.

    d) It ensures consistent output formatting across model versions.

3. What is the primary benefit of Modular prompting for reasoning tasks?

    a) It increases randomness in output generation.

    b) It creates longer outputs, which are easier to evaluate.

    c) It simplifies reasoning by isolating subcomponents of a complex task.

    d) It avoids the need to specify role or tone in each step.

4. How does Socratic prompting contribute to ethical problem-solving with LLMs?

   a) By encouraging the model to provide legal advice directly

   b) By instructing the model to avoid controversial topics

   c) By guiding users through a sequence of reflective questions without asserting conclusions

   d) By rephrasing all answers into passive voice for neutrality

5. What makes reflective prompting an ethically distinct application of LLMs in personal contexts?

   a) It bypasses the need for the CASTROFF framework.

   b) It relies solely on predefined templates.

   c) It positions the model as a thinking partner to support user agency and self-awareness.

   d) It encourages users to seek quick answers to life dilemmas.

CHAPTER 11

Prompting in Real-World Systems

11.1 What Makes a Prompt "Systemic"?

In the early chapters of this book, we approached prompt engineering as a practice rooted in direct interaction: a human writes an instruction, and a language model responds. This framing positions prompting as a discrete, purposeful act of communication between an individual and an AI system. However, in real-world applications, prompts rarely exist in isolation. They are not merely textual inputs issued through a chat window, but are increasingly embedded into tools, services, and infrastructures, forming part of larger sociotechnical systems. Within these systems, prompts shape interactions invisibly. The user may not type a command or even see the prompt, yet they remain subject to its design. It is in this context that prompting becomes systemic.

Systemic prompts operate within larger operational ecosystems, such as healthcare chatbots, learning management systems, or document summarization tools, where interface actions automatically trigger instructions without requiring direct user input. Instead of asking a model explicitly to "summarize this email," the user clicks a button, selects a preset tone, or initiates a workflow. Behind these actions lies a pre-authored prompt, constructed in advance by an engineer, designer,

H. Tavakoli, *Prompt Engineering for Everyone*,
https://doi.org/10.1007/979-8-8688-2338-1_11

or product team. The system automatically converts user actions into instructions while obscuring their underlying content, assumptions, and tone. Consequently, systemic prompts not only produce outputs but also actively mediate experiences, steer behavior, and control knowledge dissemination.

The implications of this shift are substantial. When prompts operate as invisible infrastructure, they carry responsibilities that far exceed the boundaries of a single interaction. A summarization prompt embedded in an email client may influence how professionals understand colleagues' messages across thousands of workplaces. A translation prompt in a public service kiosk may determine the accessibility of vital legal or medical information. A rewrite prompt designed to make text "more professional" may, without visibility or consent, alter the voice of a user unfamiliar with the dominant norms of managerial English. In each case, the prompt does not merely respond; it governs.

Prompt engineers must, therefore, treat prompts as systemic artifacts rather than isolated designs, anticipating how they will perform at scale, adapt across contexts, or meet diverse user expectations and requirements. Crucially, the person affected by a systemic prompt is often not the one who authored it. The dissociation between prompt creation and prompt experience introduces a new layer of ethical responsibility. Engineers must account for this distance by designing with transparency, adaptability, and failsafes that protect the user even in the absence of direct input or visibility.

Consider, for example, a municipal planning tool that allows residents to submit feedback on new housing policies. Behind each submission form, the system may automatically trigger a prompt that rephrases the resident's message for tone or clarity before summarizing it for internal use. If this prompt is overly aggressive in its rewriting, it may dilute public dissent, shift the tone of community input, or render concerns invisible, all without the user's knowledge. What appears to be a minor act of assistance becomes, in effect, an intervention in democratic communication. In such cases, prompt engineering becomes a form of institutional design.

This section introduces the conceptual and practical transition from prompting as a text-level activity to prompting as a systemic practice. It reframes prompts as infrastructural elements within platforms and workflows, components that carry authority, encode assumptions, and shape user experience at scale. As this chapter demonstrates, treating prompts systemically compels us to design not only for clarity or correctness, but for accountability, inclusion, and long-term reliability in the systems we help build. Across real deployments, systemic prompts move through a simple lifecycle of design, deployment, monitoring, revision, and retirement.

11.2 Prompting Through Tools: Interfaces and No-Code Workflows

In everyday use, most people never see a prompt, let alone write one. They do not experiment with temperature settings, issue multi-step reasoning commands, or compose carefully structured instructions. Instead, they interact with applications, dashboards, and automated services. These tools embed prompts within their interfaces, translating user actions, such as clicking a button, uploading a document, or choosing a tone, into instructions for a language model. In these contexts, prompting becomes invisible, yet no less powerful. It shifts from a human-authored message to a system-triggered function. This shift transforms both the nature of the prompt and the ethical responsibilities of those who design it.

The rise of no-code and low-code platforms has accelerated this trend. Educators, healthcare workers, content creators, and service professionals, many of whom have little to no programming experience, now deploy AI-powered features through drag-and-drop interfaces and pre-built templates. A button labelled *"Rewrite in Plain Language"* may rely on a prompt written months earlier by a product team, never reviewed by its users, and untraceable to its origin once deployed. The prompt does not merely determine the language of the output; it governs the interaction between the user and the system. It defines what counts as “plain,” “professional,” or “neutral,” thereby embedding a specific worldview into a user’s communication.

From a prompt engineering perspective, this evolution demands a new design mindset. Engineers are no longer writing for themselves or expert users. Prompt engineers design instructions for indirect audiences who frequently remain unaware that they are interacting with AI-generated guidance. As a result, the stakes of every word increase. A minor phrasing choice, such as “make this sound more confident,” may subtly alter how a user expresses themselves, especially when repeated across different contexts. In some environments, such adjustments can enhance clarity and tone. In others, they may suppress individuality, reinforce dominant norms, or marginalize voices that do not conform to implicit expectations of professionalism.

Deploying prompts across diverse populations significantly amplifies these risks. A tone adjustment feature embedded in a career advice tool may reflect culturally coded ideas of assertiveness or politeness, favoring one communication style over others. A translation prompt may prioritize Western idioms while distorting local meaning. A summarization prompt in a classroom dashboard may standardize student performance in ways that overlook learning diversity. In each of these cases, the prompt operates silently but with real consequence, influencing how users are represented, judged, or supported.

Designing for such environments requires more than linguistic fluency. It requires a profound understanding of interface dynamics, user experience, and inclusivity. Prompt engineers must predict both the model's outputs and users' interpretations of them, particularly when individuals cannot access or modify the underlying instructions. This real-world challenge impacts crucial systems, including legal aid bots for immigrants, health guidance platforms for elderly users, and educational content moderation tools. In these systems, the prompt must serve a broad spectrum of users, many of whom rely on it in moments of need, confusion, or vulnerability.

Moreover, prompts in interface-based tools are rarely static. They are updated, reused, and redeployed across features and products without consistent review and oversight. Without transparent processes for versioning, feedback, and user consultation, these prompts risk becoming invisible factors that determine the digital experience. The model's response may appear seamless, but its logic may rest on outdated, untested, or exclusionary assumptions. A lack of transparency reinforces opacity and erodes user agency.

This section invites prompt engineers to rethink the role of prompting in no-code and interface-driven environments. The prompt is not merely a technical input. It is a hidden mediator of interaction and meaning. As systems shift from user-authored instructions to designer-written templates, the responsibility burden grows. Prompt engineers must adopt a dual role: part interface designer, part language ethicist. They must write with empathy, anticipate consequences, and ensure that every embedded instruction respects the autonomy, identity, and diversity of its users.

11.3 System Failures and Feedback Loops

In real-world deployments, prompts do not operate in isolation. Once embedded within tools, services, or workflows, they become active components of larger systems, systems that must function consistently, equitably, and safely over time. A single correct output cannot measure the effectiveness of a prompt in these settings. Evaluations must assess sustained reliability, diverse user impacts, and adaptive responsiveness, as prompts that succeed in testing often fail when encountering new user demographics, evolving contexts, or subtle shifts in usage patterns.

Systemic prompt failures rarely announce themselves. They are often silent and cumulative. Rather than producing obvious errors or offensive content, they tend to degrade user experience gradually. A prompt that instructs a model to "respond helpfully" may work well for one demographic but yield vague or irrelevant outputs for another. A customer support tool designed to "de-escalate complaints quickly" may prioritize brevity over empathy, thereby suppressing legitimate user concerns. A health explainer that simplifies complex diagnoses may consistently downplay symptoms more commonly reported by specific populations. In each case, the prompt has technically succeeded in generating text, but functionally failed in serving its intended purpose.

Such failures often stem not from malice or negligence, but from the absence of structured feedback. Deploying prompts without monitoring mechanisms allows discrepancies between intended effects and actual outcomes to persist undetected across different contexts and over time. These mismatches accumulate, undermining trust and perpetuating structural bias. Because many real-world systems do not expose the underlying prompt to users, or even to administrators, diagnosing the source of failure becomes difficult. A user may disengage, misinterpret an answer, or make a poor decision, all without knowing that an invisible prompt shaped the outcome.

To address this challenge, prompt engineering must integrate feedback loops as a core design practice. Effective feedback loops first establish observability, monitoring, and prompt performance in live environments by tracking outputs, user behavior, satisfaction metrics, retention rates, and complaint trends. Observability allows engineers to distinguish between isolated anomalies and recurring problems. A slight decline in response quality across several user groups may indicate a more significant issue with how the prompt is interpreted or applied.

Once patterns of failure are detected, systems must support traceable revision. Unlike ad hoc prompt editing, systemic prompts require documented version control. Each change, however small, should be recorded along with its rationale and contextual performance data. Documenting phrasing rationales maintains ethical accountability while enabling future engineers to reconstruct the decision-making process behind confident language choices. More importantly, it enables structured experimentation. A/B testing of alternative prompts, for example, can reveal which phrasing better supports inclusion, clarity, or user trust.

Prompt revision is not solely a technical task. For sensitive applications such as education, mental health support, or legal assistance, domain experts must collaboratively assess and make prompt modifications to verify their effectiveness and appropriateness. A new summary style may appear more concise, but it could also omit critical nuances. A prompt intended to reduce ambiguity may inadvertently suppress emotional tone. These trade-offs require thoughtful deliberation and interdisciplinary input. Feedback loops must therefore include not only system metrics but also human perspectives, especially from those communities most affected by the outcomes of the prompt.

Crucially, systems must allow for human oversight. In high-risk contexts, no prompt should operate without a clear pathway for escalation. If a model detects a potential crisis, such as suicidal ideation, legal jeopardy, or abuse, it must defer to a human responder, offer appropriate support options, or suspend interaction altogether. These safeguards mark the boundary at which automated responses must stop and a human decision-maker assumes responsibility. Confidence thresholds and moderation checkpoints are not optional enhancements; they are essential components of the system. They are foundational safeguards that distinguish responsible deployment from reckless automation.

Viewed in this light, prompts are not static instructions. They are iterative design artifacts with lives of their own. They must be governed, versioned, evaluated, and, when necessary, retired. The goal of a feedback loop is not to reach a final, perfect prompt but to sustain a cycle of continuous improvement grounded in real-world use.

Prompt engineers working in live systems must adopt a new professional identity: part designer, part ethicist, part system steward. Their responsibility is not only to craft linguistically sound instructions but to build and maintain the structures that ensure those instructions serve users fairly over time. Without these feedback architectures, even the most elegant prompts risk becoming brittle, unaccountable, and harmful in practice.

The accurate measure of a prompt is not the elegance of its wording but the quality of the human experiences it shapes. Feedback loops are how we measure, improve, and honor that responsibility.

11.4 Design Patterns and Deployment Ethics

As prompts become embedded in tools, platforms, and public services, their function changes. What begins as a single line of instruction transforms into a mechanism that governs interaction, access, and even institutional tone. A prompt embedded in a system is no longer a private act of communication between a user and a model; instead, it becomes a public act of communication between a user and a model. It is a public-facing design decision, one that may persist across thousands of interactions and shape outcomes in ways the original writer never anticipated.

This section identifies four recurring design patterns in the deployment of prompted systems: *chat-based assistants, background generators, agent pipelines, and simulation-based prompts*. Each of these patterns introduces distinct affordances and risks. Understanding them is essential to recognizing how systemic prompts shape the user experience, especially when their design is invisible to the end user.

*Chat-based assistant*s, the most prevalent pattern, simulate dialogue across various applications, including education, customer support,

HR, and coaching. While presenting as conversational, these systems operate under strict system-level prompts that govern tone, boundaries, and content parameters. These role-defining prompts are rarely visible to users but strongly influence the authority, empathy, and scope of responses. When deployed in sensitive contexts, such as legal advice, health triage, or student guidance, these assistants may convey undue confidence or false expertise. The risk lies not just in hallucinated content, but in the impression that the model speaks with institutional authority. Effective prompt design for such systems requires explicit boundaries, unambiguous disclaimers, and calibrated tones that communicate reliability limitations, thereby preventing users from misinterpreting linguistic fluency as substantive credibility.

The second pattern is the *background generator*. These systems offer features such as automatic summarization, tone rewriting, translation, or content completion, often triggered by a click or contextual cue. The system designer's predetermined linguistic framing, invisible to end users, often embeds unexamined biases within these concealed prompts. A button labelled "Make more professional" might consistently favor corporate idioms, Western politeness norms, or formal sentence structures, regardless of context. Over time, such prompts can homogenize language use and subtly erode the user's voice. The ethical question here is not whether the system works, but whose standards it enforces. Prompt engineers must recognize that they govern not merely textual outputs but modes of expression, effectively controlling how users may present themselves within the system

The third paradigm, *agent pipelines*, sequences multiple prompts to simulate planning, execution, and assessment cycles, frequently applied in research synthesis, process automation, and code generation contexts.

While powerful, their complexity can obscure accountability. A flaw in the initial task decomposition may propagate through every subsequent step, producing outputs that appear coherent but are grounded in faulty assumptions. Because users often see only the final product, they have little insight into where errors originated or how to intervene. Designing these pipelines requires not only technical precision but also modular transparency and clarity. Prompts at each stage must be traceable, auditable, and revisable. Moreover, the system should offer users a way to inspect intermediate steps or request human review when confidence is low. Without this, agent pipelines risk becoming black boxes that mask flawed reasoning under the illusion of procedural logic.

The final pattern is *simulation-based prompting*. These systems direct models to adopt specific perspectives, whether those of historical figures, whistleblowers, or community elders, to simulate dialogues, training exercises, or ethical quandaries. While this enables rich learning environments, it also raises critical concerns around representation and authenticity. When a prompt instructs the model to "act like a refugee mother" or "simulate an indigenous activist," the resulting portrayal is shaped by the training data, which may draw on media stereotypes or incomplete cultural narratives. These simulations can unintentionally reinforce caricatures or reproduce exclusionary worldviews. Prompt engineers must critically examine which perspectives their systems simulate and amplify. On what basis? And with what constraints? A simulation prompt must communicate that it is an approximation, not an authentic embodiment of lived experience. It must also offer respectful, grounded representations that avoid reducing complex identities to simplified archetypes.

Across all four patterns, a central ethical theme emerges: *prompts are not neutral*. They encode decisions about tone, form, knowledge boundaries, and user positioning. They shape not only what users can ask but also what they are likely to receive, believe, or act upon. For this reason, the role of the prompt engineer extends far beyond linguistic tuning. It becomes an ethical and institutional responsibility. Each deployment requires reflection not only on how the system functions but also on whom it serves, under what assumptions, and what effect it has.

Large-scale prompt deployments necessitate rigorous design reviews incorporating transparency standards, feedback systems, and equity evaluations. Systems must support challenge and revision, not only by developers but also by those communities most affected by the outcomes. Where users cannot see or contest a prompt, engineers must build safeguards into the workflow itself. These include clear disclaimers, user-controlled adjustments, or escalation pathways to human support.

The professionalization of prompt engineering depends on the maturity of these practices. The prompt, once a simple input, now functions as a policy instrument, a cultural filter, and a social contract. Its phrasing, positioning, and visibility can influence dignity, clarity, and trust at scale. Treating prompts as design artifacts, rather than invisible infrastructure, is essential for ensuring that AI systems remain responsive not just to inputs, but to the values and expectations of those they serve.

11.5 Governance Interfaces for Prompted Systems

In earlier chapters, prompts were treated primarily as linguistic artifacts: carefully constructed inputs that guided model behavior in discrete exchanges. However, once embedded in real-world systems, prompts assume a different status. They no longer operate as ephemeral instructions but as infrastructural components woven into applications, services, and organizational workflows. At this scale, a prompt is not simply a query; it becomes a durable design choice with the power to shape institutional behavior, influence user trust, and mediate experience over time. Such permanence demands governance frameworks that extend beyond design creativity toward sustained oversight and accountability.

Prompt governance refers to the institutional and technical practices that ensure prompts remain safe, transparent, and auditable throughout their lifecycle. Every deployed prompt functions as a governance artifact, requiring documentation of authorship, intent, safeguards, and review cycles. This is especially crucial in high-stakes settings, such as healthcare, employment, education, or the justice system, where even subtle wording can impact fairness, legal liability, or individual well-being.

Effective governance rests on four core capacities: traceability, auditability, version control, and shared accountability. Traceability provides records of authorship, rationale, and approval history, enabling investigators to reconstruct the reasoning behind a prompt's specific design. Auditability extends this by allowing both internal and external stakeholders to test how prompts shape outputs, including whether they introduce systematic bias. *Version control* documents every phrasing change, enabling teams to explain, test, or roll back shifts in model behavior. *Shared accountability* recognizes that, unlike experimental prompting, where a single user bears responsibility, institutional prompting distributes responsibility across designers, domain experts, legal advisors, and vendors, while end users may remain unaware of the wording that governs their interaction.

To operationalize these principles, organizations benefit from a **System Prompt Register**, a structured inventory that captures the essential governance details of each deployed prompt. A standard template might include

- **Location:** Where the prompt is embedded in the system or workflow
- **Trigger:** The event or input that activates the prompt
- **Owner:** The team or individual responsible for its maintenance
- **Purpose:** The communicative or operational aim of the prompt
- **CASTROFF fields:** Which dimensions are being constrained (e.g., audience, scope, tone)

- **Data flows:** What information is passed into or out of the prompt
- **Last review date:** The most recent formal audit

Such a register provides visibility across teams, strengthens accountability, and supports regulatory compliance. A short revision example later in the text shows how an embedded prompt can be updated through this register rather than rewritten ad hoc.

Visibility also requires design variation. In some contexts, prompts should be exposed directly as visible fields for power users, allowing them to adjust tone, scope, or format with precision. In others, especially public-facing systems, guardrail prompts with progressive disclosure are preferable: the user interacts with a simplified surface, while deeper instructions remain managed by the system and are disclosed only when necessary for clarity or consent. This dual pattern strikes a balance between flexibility for experts and safety for general users.

Robust governance further depends on change-control mechanisms. Prompts must be versioned like code, with documented release histories. Controlled A/B tests measure the impact of changes before broad deployment. Rollback plans are essential in case a new prompt produces unintended harm, while audit hooks ensure that each update leaves an examinable trail. Without such mechanisms, even minor edits risk destabilizing user trust or regulatory compliance.

These practices highlight that prompt governance is neither ornamental nor optional. Dashboards that track authorship, versioning, approval status, and performance metrics need not be elaborate, but they are indispensable. Absent this infrastructure, even skilled practitioners risk becoming invisible authors of systemic harm, unable to explain or correct the outputs they helped create.

This section, therefore, reframes prompt engineering as a sustained professional practice rather than a creative exercise. Prompts are not static artifacts but evolving governance objects whose design, deployment, and revision require collective oversight. The following sections extend this discussion into multilingual and cross-cultural contexts; yet, the principles begin here, with the infrastructure of registers, patterns, and change-control protocols that make prompt design as accountable as it is expressive.

11.6 The Urgency of Cross-Cultural and Multilingual Prompting

As artificial intelligence systems continue to expand globally, their capacity to serve a truly diverse population remains critically uneven. Public rhetoric touts AI's global reach, yet linguistic disparities persist. Models overwhelmingly favor English and a few other high-resource languages, creating structural barriers that undermine access, trust, and equity in practical applications. In this context, multilingual and cross-cultural prompt engineering is not a secondary concern; it is a crucial aspect of the process. It is a human-centered imperative.

Language in AI systems is not a neutral conduit. It is a cultural vessel that carries values, norms, idioms, hierarchies, and histories. English-centric prompt design embeds that language's cultural assumptions, systematically influencing tone, politeness conventions, and expressions of authority, risk, and emotion across all outputs. For users in non-English-speaking regions or those who speak English as a second language, these embedded norms may feel unfamiliar, exclusionary, or even disrespectful. This gap becomes especially dangerous in high-stakes domains such as healthcare, law, education, and public services, where precision and empathy are essential.

The training methodologies for language models fundamentally create multilingual prompting challenges. The underlying datasets used to pre-train large models draw heavily from the internet, where English is the dominant language. Current multilingual fine-tuning implementations demonstrate inconsistent scope, variable quality, and patchy linguistic coverage despite recent advancements. As a result, prompts written in many global languages produce outputs that may be fluent on the surface but culturally or semantically misaligned beneath. A model may respond in grammatically correct Swahili or Hindi but still miss idiomatic nuance, emotional cues, or contextually appropriate expressions. In some cases, it may default to English logic entirely, translating metaphors, professional norms, or argument structures in ways that distort the original intent.

This issue intensifies when systems treat linguistic generalization as inconsequential. A prompt labelled *make this sound more formal* may generate acceptable results in English, where formality is often associated with indirectness or passive voice. However, in Japanese or Arabic, formality may require entirely different sentence structures, honorifics, or culturally specific conventions. Without deep linguistic knowledge embedded in the system, such prompts produce misleading or even disrespectful results.

The stakes are high. Consider a public health chatbot deployed in Brazil that misinterprets a user's tone in Portuguese and fails to detect emotional distress. Alternatively, a legal guidance tool that processes Bangla inputs using an English-anchored reasoning structure, thus simplifying or misrepresenting the protections available under local law. These are not hypothetical scenarios. Early deployments across the Global South already reveal these exclusionary patterns, demonstrating how linguistically inflexible AI systems actively marginalize users.

The work of multilingual prompt engineering, therefore, is not limited to translation. Effective localization requires adapting prompts to align with the specific cultural norms and linguistic expectations of a particular community, meticulously addressing tone, idioms, formality levels, and conceptual frameworks. A prompt written for a Ghanaian farmer using Twi should not simply be a translated version of an English agricultural tip. It must reflect local farming knowledge, seasonal metaphors, and communication norms rooted in lived experience. Similarly, a mental health support prompt for a Korean user must account for culturally specific expressions of stress, shame, and emotional restraint. The goal is not just fluency. It is relevance, accuracy, and respect.

However, this responsibility cannot fall solely on prompt engineers. Institutions, developers, and policymakers must recognize linguistic equity as a foundational principle of AI deployment. Substantial investment in language-specific training corpora, grassroots collaborations, and regionally adapted evaluation frameworks becomes essential. Without this foundational infrastructure, even meticulously crafted prompts fail to deliver consistent cross-linguistic performance.

Prompt engineering, when viewed through a multilingual lens, becomes a profoundly ethical practice. This approach critically examines the voices the system incorporates, the linguistic patterns it recognizes, and whose experiences shape its underlying logic. In a world of over 7,000 living languages, designing for only a few is not a technical necessity; it is a choice. Moreover, that choice has consequences.

If prompt engineering is to become a globally relevant profession, it must adopt the challenge of multilingual design not as an afterthought, but as a core competency. A monolingual paradigm cannot fulfil the promise of AI. Only by embracing linguistic diversity at the level of prompt design can we begin to build systems that truly serve the full spectrum of humanity.

11.7 Equity-Driven Solutions for Multilingual Prompting

The disparities outlined in the previous section are not incidental technical gaps. They are structural outcomes of how language models are trained, deployed, and governed. Most AI systems today perform best in English and a few dominant world languages, while users who interact in low-resource or non-Western languages experience degraded, distorted, or culturally misaligned responses. These inequities do not merely inconvenience users; they reproduce digital marginalization and restrict access to essential services, knowledge, and opportunities. If prompt engineering is to be both globally relevant and ethically grounded, it must embrace solutions that are explicitly equity-driven.

Addressing linguistic disparities requires a shift in design priorities. AI systems require multilingual foundations from inception rather than English-first designs with subsequent translations. This transformation must start with training data, currently dominated by English-language digital content that distorts models' linguistic capabilities. Correcting this imbalance involves the deliberate inclusion of underrepresented languages, not just in terms of quantity but also with attention to quality, diversity, and domain relevance. This process must involve linguists, educators, cultural historians, and community experts who understand not only grammar but also the lived realities that those languages encode.

Accurate linguistic equity demands systemic changes across the entire prompt lifecycle, from design and testing to deployment environments, rather than relying solely on model training adjustments. Many current tools allow inputs in different languages but fail to support complete interaction cycles. Even when models respond fluently in Spanish or Bengali, English-dominated interfaces, instructions, and error messages can fracture the user experience, discouraging experimentation and perpetuating reliance on English-speaking mediators. Designers must treat localization as a primary goal, not a finishing touch. Interfaces should support tone selection, task explanations, and contextual help in users' preferred languages, configured not only for fluency but also for cultural fit.

Global institutions must equitably distribute the capabilities to develop, regulate, and scrutinize multilingual prompting systems. Many countries that would benefit most from inclusive AI systems currently lack access to training, datasets, research tools, or prompt engineering expertise in their national languages. To address this, governments, universities, and civil society organizations must invest in language-specific hubs that support research, prompt libraries, evaluation frameworks, and documentation tailored to local needs. In contexts where this capacity is still developing, international collaboration is essential. Cross-border partnerships can bootstrap local ecosystems by sharing infrastructure, co-developing resources, and recognizing local knowledge as essential to the global AI commons.

Global organizations, such as the United Nations, UNESCO, and the World Bank, also play a critical role. Their mandates, focused on development, education, and cultural preservation, extend naturally into the domain of equitable AI. These institutions have the reach and credibility to promote multilingual AI as a public good, set inclusive standards, fund research into low-resource language technologies, and require linguistic transparency in public sector deployments. Prompt engineering, when shaped by such efforts, becomes not only a design task but a vehicle for rights-based inclusion.

The ethical case is clear: people should not have to switch languages to access AI that works. Nor should they be asked to abandon local ways of speaking in favor of imported norms encoded in another language's grammar, idioms, or professional tone. Language is not simply a conduit for content. It is a carrier of identity, dignity, and local knowledge. Designing prompts that respect this fact is not a matter of convenience; it is a matter of respect. It is a matter of justice.

Equity-driven prompting necessitates designing systems that adapt to users' linguistic, cultural, and social contexts, demanding prompt engineers to reject linguistic defaults, reassess traditional clarity norms, and actively incorporate historically marginalized perspectives into technical frameworks. It also requires institutional willingness to fund, govern, and sustain this work as core infrastructure, not as an optional enhancement.

The future of prompt engineering is either multilingual or exclusionary by design. If we are to realize the full potential of AI as a tool for human empowerment, we must recognize linguistic equity as a central pillar of system design. A prompt is not just a string of words. It is a choice about whose world the system understands and whose voice it is prepared to hear.

11.8 Summary

This chapter marked a turning point in our understanding of prompt engineering by shifting the focus from standalone instructions to prompts as integral components of deployed systems. While earlier chapters established prompting as a communicative craft between a user and a model, this chapter recast it as a form of system *design*, one that operates at scale, shapes institutional behavior, and quietly governs user experience. The result is a redefinition of the prompt engineer's role: from linguistic technician to ethical architect of sociotechnical infrastructure.

We began by exploring the concept of systemic *prompting*. Systemic prompts integrate directly into tools, platforms, and operational workflows rather than functioning as standalone user inputs. These prompts are often invisible to users, triggered by interface actions or default settings rather than written text. Systemic prompts integrate directly into tools, platforms, and operational workflows rather than functioning as standalone user inputs. When prompts become systemic, their ethical stakes increase. They must be designed not only for accuracy and efficiency but also for fairness, adaptability, and long-term accountability.

In Section 11.2, we examined how prompts operate through interfaces, particularly in no-code and low-code environments. Most users today interact with AI through applications, not by writing instructions. Engineering and product teams systematically orchestrate these prompts by predetermining their tone, structural framework, and intentional design. While this makes AI more accessible, it also increases the risk of invisible bias and cultural homogenization. Prompts embedded in buttons such as *"Make more professional" or "Summarize this email"* encode assumptions that may marginalize specific ways of speaking or communicating. Prompt engineers in these systems must therefore think as interface designers, anticipating the needs of unseen users and designing with empathy, inclusivity, and transparency.

The chapter then turned to the issue of systemic *failure*. Systemically embedded prompts often reveal their flaws through subtle, cumulative effects rather than conspicuous failures. More often, they appear as quiet degradations in user experience, responses that misalign with context, reinforce stereotypes, or fail to support vulnerable users.

Section 11.3 argued that robust prompt engineering must include built-in feedback *loops*. These involve observability, version control, structured testing, and the capacity for human oversight. A prompt is not a static command; it is an evolving design artifact. Sustaining its quality requires continuous monitoring, interdisciplinary review, and mechanisms for timely correction.

In Section 11.4, we analyzed four real-world design *patterns* in prompt deployment: chat-based assistants, background generators, agent pipelines, and simulation-based systems. Each pattern offers specific benefits and introduces distinct ethical risks. A simulation prompt may foster empathy, but it also risks caricature. A summarization tool may improve efficiency while erasing nuance. These patterns demonstrate that prompts are never neutral; they carry values, voice, and authority. They must therefore be subject to ethical design review and transparent governance.

The following section focused on linguistic *equity*. Section 11.6 exposed the global asymmetry in language model performance, showing how most AI systems serve English speakers far better than those using low-resource or non-Western languages.

Section 11.7 followed with a clear call to action: multilingual support is not an afterthought but a fundamental requirement of ethical design. Prompt engineering must actively resist linguistic exclusion by supporting native-language interfaces, culturally grounded design, and the localization of tools across all regions and populations. The work of building equitable AI begins with recognizing language not just as a feature, but as a form of access, identity, and justice.

In sum, this chapter redefined the boundaries of prompt engineering. It is no longer simply about what a prompt tells a model to do. It is about what that prompt makes possible for others, on a larger scale, over time, and across differences. In real-world systems, prompting is no longer an isolated act; it is an integral part of the overall process. It is a public responsibility.

11.9 Self-Assessment Quiz

1. How does the shift from standalone prompts to systemic prompting change the ethical responsibilities of prompt engineers?

   a) It reduces the need for careful phrasing since prompts are reused across contexts.

   b) It increases the stakes by making prompts affect large-scale user experiences and institutional outcomes.

   c) It allows users more control, thereby reducing the engineer's accountability.

   d) It ensures that errors are automatically corrected through system updates.

2. What is one risk of prompts embedded in no-code interfaces that users never see?

    a) Users might accidentally rewrite the system prompt.

    b) Prompts may conflict with system memory settings.

    c) Invisible prompts can enforce biased or exclusionary standards without the user's awareness.

    d) These prompts always default to generic instructions and cause hallucinations.

3. Why are feedback loops essential in systems that use prompted AI at scale?

    a) They allow faster responses during peak traffic.

    b) They allow users to skip the prompt design altogether.

    c) They provide a method for observing prompt performance, identifying failures, and supporting revision.

    d) They ensure that prompts never produce repetitive content.

4. In agent pipeline systems, what is a key concern when prompts are chained together?

    a) The system becomes too easy to audit.

    b) Errors in early prompts may propagate invisibly through each stage.

    c) The prompts fail to run when multiple languages are used.

    d) Users can edit prompts mid-pipeline, causing system instability.

5. Why is multilingual prompt engineering essential for equitable AI deployment?

    a) It enables more creative outputs in English.

    b) It reduces computing costs during fine-tuning.

    c) It ensures that people can interact with AI in their preferred language, preserving access and identity.

    d) It allows every user to receive outputs in machine code.

CHAPTER 12

The Future of Prompt Engineering

12.1 From Skill to Discipline: What Prompt Engineers Are Becoming

When large language models first entered the public domain, prompting appeared improvisational and organic. Early users approached it as a craft or game, experimenting with phrasing, adjusting tone, and employing clever techniques to produce novel results. Online communities shared "prompt recipes" and exchanged methods for eliciting creative or functional responses. In this initial phase, prompting resembled digital tinkering: playful, personal, and informal. However, this era is drawing to a close.

Today, prompting has evolved into a formal design responsibility. No longer merely a novelty or workaround, it now constitutes a foundational layer of applied AI. Across education, healthcare, governance, creative industries, and public services, prompt engineering directly shapes how language models function in real-world systems. The individual who writes the prompt determines the information surfaced, frames decisions, and influences user experience; prompt engineers, once informal explorers, now architect institutional logic.

H. Tavakoli, *Prompt Engineering for Everyone*,
https://doi.org/10.1007/979-8-8688-2338-1_12

This section examines the transformation of prompting from a flexible skill into an emerging professional discipline. The shift is not only technical but also conceptual and ethical in nature. A discipline is defined not only by its methods but by its standards, literacies, and responsibilities. Prompt engineering is now entering this domain.

12.2 From Play to Practice

Earlier chapters described prompting as a layered form of instruction, requiring structure (see Chapter 10), contextual sensitivity (see Chapter 5), and cultural awareness (see Chapter 11). Initially, prompting focused on tactics, including brief queries, inventive phrasing, or single-turn experiments, designed to elicit specific outputs. Users viewed prompts as individual tools, often crafted in isolation and evaluated by immediate results. This perspective no longer aligns with current realities.

Modern deployments embed prompts within platforms, tools, and services that reach thousands of users. A prompt written for a legal chatbot or mental health assistant can influence how individuals understand their rights or seek care. Prompts now direct how systems communicate, summarize, advise, and respond. Their impact extends beyond isolated conversations, shaping public experience at scale.

As a result, prompt engineering is now a *design discipline rather than a mere technical trick*. Prompt engineers must balance technical clarity with social responsibility. Their role requires a comprehensive understanding of communication, ethics, and the underlying human needs in every interaction.

To clarify this evolution, the table below contrasts the *early view of prompting* with the *emerging professional role*:

Early View	Evolving Role Today
Write clever one-liners	Design clear, goal-aligned, and ethical instructions
Explore model quirks	Understand user goals, risks, and system context
Optimize for novelty or surprise	Optimize for clarity, safety, and meaningful outcomes
Work in isolation	Collaborate across design, policy, and domain-specific teams
Treat the model as a puzzle	Treat the model as a communicative tool with real-world impact

12.3 The Literacies of a Discipline

Every profession is grounded in shared literacies: distinct ways of thinking, acting, and caring. Based on the preceding chapters, four core literacies shape the maturation of prompt engineering.

1. **Cognitive Literacy**

   Prompt engineers must be able to think in structured steps, break down tasks, and scaffold outputs. As seen in Chapter 10, strong prompts are not casual commands; they are instructional frameworks that guide reasoning and manage complexity.

2. **Ethical Literacy**

   As Chapter 11 showed, prompting is never neutral. The words we choose can reinforce bias, exclude perspectives, or affect user dignity. Ethical literacy involves learning to anticipate potential harm, question underlying assumptions, and design solutions with care.

3. **Communication Literacy**

   Prompts are dual-addressed: they speak to the model, but their effects shape human experience. Prompt engineers must be fluent in tone, clarity, and cultural nuance, especially in multilingual and multicultural settings.

4. **Systemic Literacy**

   Prompts are increasingly operating within platforms, institutions, and automated workflows. Engineers must learn to design within these systems, understanding compliance needs, accessibility requirements, and long-term maintenance structures.

12.4 Rethinking Who Counts As a Prompt Engineer

Some readers may hesitate to identify as "prompt engineers," particularly if they do not work in software companies or write code. However, this book adopts a broader perspective: anyone who writes prompts that affect others, especially in systems of care, education, communication, or service, is practicing prompt engineering.

Job titles do not define professional identity, but rather, intent, care, and responsibility. Prompt engineering is becoming a discipline not due to formal qualifications, but because it requires principled practice.

This discipline is built not through certification, but through sustained commitment. Prompt by prompt, decision by decision, practitioners shape a field that blends human understanding with technological precision, thereby defining how future systems will communicate, serve, and support those they reach.

12.5 The Expanding Prompt Ecosystem

For many, the phrase "prompt engineering" *still* evokes a simple image: someone typing a question into a chat window and receiving a response from an AI. While this scenario remains common, it no longer reflects the full scope of how prompting functions in contemporary systems. In practice, prompts now extend far beyond direct human-to-AI interaction. They operate across a broad ecosystem of tools, platforms, and institutional environments, often invisibly, yet with substantial influence.

Earlier chapters framed prompting as a human-centered act of language design, creative, structured, and ethically attuned. While this framing remains essential, the context has evolved. Prompt engineers now integrate prompts into medical support applications, classroom dashboards, legal summarization systems, corporate knowledge bases, and civic service platforms. Teams rather than individuals increasingly author prompts, and organizations execute them thousands of times per day. These prompts create ripple effects that shape not only machine outputs but also human understanding, decision-making, and behavior.

To understand the future of prompt engineering, it is essential to recognize it as part of an expanding ecosystem: an interconnected network of technical and social systems in which prompting occurs continuously, at scale, and often without user awareness.

12.6 From Interaction to Infrastructure

Prompting used to be a moment of direct interaction: a user would write a command, and the model would reply. Most users remain unaware of this moment. When they click *"Summarize," "Rewrite in Plain Language,"* or *"Suggest a Headline,"* they do not issue their prompts. They are activating prewritten instructions embedded within the software. The prompt is still there, but it was authored in advance by someone else and is running silently in the background.

This transition from visible commands to invisible infrastructure introduces new design challenges. A prompt embedded in a health application must function reliably across diverse inputs and users, upholding clinical safety while remaining sensitive to emotional tone. Similarly, prompts embedded in educational tools must strike a balance between clarity and depth, accommodating learners with varying needs and abilities. Government chatbots require prompts with legal precision, cultural neutrality, and linguistic inclusivity. These are not merely language choices, but design decisions with institutional significance.

In this context, prompt engineers are no longer just crafting queries; they are also creating and refining them. They are building components of systems. Prompt engineers must adopt the mindset of infrastructure designers, determining how knowledge flows, establishing trust mechanisms, and governing how language wields power in everyday tools.

12.6.1 Institutional Contexts and Responsibilities

Consider education. In many schools, students now interact with AI-powered tutors, writing assistants, or learning dashboards. These systems generate explanations, rephrase feedback, and provide tailored support. However, the prompt behind each response must reflect more than correctness. It must also respect pedagogical values, accommodate different learning styles, and be accessible to students with disabilities. If it simplifies too aggressively, it may distort meaning. If it uses an unfamiliar tone, it may alienate learners. The responsibility here is not just technical; it is educational and ethical.

In healthcare and mental health settings, prompting takes on a different urgency. A wellness assistant responding to a distressed user must do more than sound helpful. Its prompts must convey support, avoid triggering language, and escalate to human help when needed. Misalignment here can deter people from seeking care or mislead them during moments of vulnerability. Prompt engineers must collaborate with clinicians and patient advocates to ensure the system speaks with both safety and compassion.

Legal and civic platforms present another challenge. A chatbot summarizing a tenant's rights or explaining tax obligations must be accurate, unbiased, and easy to understand across diverse audiences. The prompt that elicits such a response must be traceable, auditable, and carefully phrased. If it simplifies too much, it may confuse. Biased institutional prompts may disadvantage their intended users. In these contexts, prompt engineers steward civic clarity.

Creative industries, too, increasingly rely on prompts. Storyboarding tools, writing assistants, and marketing generators all use prompt-based instructions to determine tone, genre, and emotional rhythm. Prompt engineers in these environments must avoid clichés, preserve cultural authenticity, and ensure that generative systems amplify diverse voices rather than flattening them into stylistic conformity.

Even in everyday business tools, such as collaborative documents, productivity dashboards, and customer service inboxes, prompts operate quietly yet pervasively. Features like "Rewrite for tone," "Summarize this thread," or "Suggest reply" all rely on prompts designed by others. These systems shape communication norms, influence workplace dynamics, and affect how users present themselves professionally. The prompt underlying these outputs exerts a subtle, persistent influence.

12.7 Interdisciplinarity and Systemic Awareness

Today, prompt engineering is an interdisciplinary practice. Writing a good prompt is no longer enough; it must also be effective. Engineers must comprehend both the systems implementing these prompts and the populations they affect while addressing the accompanying ethical concerns. A prompt embedded in a healthcare interface must reflect clinical guidelines. One used in a public-facing civic tool must comply with accessibility standards and multilingual inclusivity. Education app prompts must accommodate neurodiversity, cultural differences, and core learning goals.

Consequently, prompt engineers must master not only programming and design but also the specialized terminology of law, medicine, education, media, and public administration. They must collaborate with domain experts, accessibility consultants, policy advisors, and affected communities. Prompting has evolved into an act of *social design*: not just a linguistic trick, but a form of communication that operates across various systems and institutions.

12.8 Reflection: Reframing Our Practice

For many readers, prompting began as an individual skill, a creative experiment, a productivity tool, a personal assistant. However, as this section demonstrates, it has evolved into something more: a form of *distributed professional responsibility*. Prompts now structure how people learn, make decisions, receive care, and access knowledge.

The question, then, is no longer simply what kind of prompts we can write, but rather what kinds of systems we help to build through the prompts we design.

If we write prompts that guide thousands of interactions, influence public discourse, and shape institutional behavior, then we are participating in the prompt ecosystem as professionals. That role carries obligations: to design with care, to anticipate impact, and to align our instructions with the values of the communities they touch.

The chapters that follow will examine how this responsibility relates to regulation, transparency, and professional codes of conduct. However, the foundation is already clear: prompt engineering is not just a skill. This role operates within systems that produce tangible real-world impacts, demanding the same rigorous standards, professional seriousness, and collective ethics as any established discipline.

12.9 Risks, Regulations, and Responsibility

As prompt engineering becomes a core practice in AI system design, the ethical stakes deepen. Prompts are increasingly influencing how people receive healthcare advice, understand their legal rights, navigate public services, and interpret educational or financial information. With each deployment, the weight of the prompt engineer's decisions grows. What practitioners once viewed as a clever technique now demands serious design accountability with measurable, real-world impacts.

Importantly, most of the risks associated with prompting do not arise from malice or misuse. Instead, they stem from omission, oversight, or insufficient system-level thinking. A prompt might mislead users because it oversimplifies an issue. It may reinforce bias because it fails to consider who is speaking and to whom they are addressing. It may unintentionally exclude users by assuming familiarity with technical language, institutional norms, or dominant cultural references. Neglect rather than intent often causes these harms, making thoughtful governance and professional accountability crucial components of modern prompt engineering.

12.10 Types of Risk in Prompting Systems

Prompt-related risks fall into three primary categories: *representational*, *epistemic*, and *institutional*.

Representational risks occur when prompts unintentionally encode or reproduce social bias, cultural erasure, or stereotypes. For instance, a customer service prompt that defines "professional tone" in ways aligned with Western corporate culture may suppress authentic expression for users from other linguistic or cultural backgrounds. Similarly, prompts that simulate personas, such as patients, parents, or community leaders, may rely on incomplete or biased training data, resulting in distorted representations of lived experiences.

Epistemic risks involve the model's knowledge claims. Prompts that request summaries, insights, or explanations can result in confident-sounding responses that are, in fact, speculative or misleading. Without safeguards, such as source citations, uncertainty indicators, or scoped instructions, language models may produce hallucinated claims that sound credible but are unfounded. In fields such as law, medicine, or science, where factual accuracy is paramount, the cost of such failures can be high. Each of these risk categories can be mitigated through scoped constraints, uncertainty statements, traceable revision, and human-in-the-loop escalation.

Institutional risks arise when prompts are embedded in operational systems without adequate attention to transparency, oversight, and accountability measures. Opaque prompt systems that prevent users from examining or contesting their influence, and organizations from tracking their development or rationale, create accountability gaps that erode institutional trust.

These risks are not abstract. They are already materializing across education platforms, legal bots, civic tools, and healthcare assistants. The key to mitigation lies not only in the prompt design itself but also in the adoption of ethical infrastructures surrounding prompting.

12.11 Regulatory Responses and Governance Frameworks

Governments and regulators are beginning to address the expanding societal impact of language models. The European Union's AI Act (2024) offers one of the most comprehensive frameworks to date. It classifies AI applications into risk categories and imposes strict requirements on "high-risk" systems, such as those used in employment, education, healthcare, or law enforcement. These requirements include human oversight, documentation protocols, and the traceability of system logic, including the role of embedded prompts.

Within such frameworks, prompt engineering becomes a strategic responsibility. The field now recognizes prompts not as transient text inputs, but as programmable policy components that govern how AI systems interpret, communicate, and interact with humans. Consequently, prompts must be documented, auditable, and designed for accountability.

At the industry level, several voluntary initiatives are emerging. Model cards, initially designed to describe the capabilities and limitations of language models, are evolving into system cards, which document not only the model but also prompt configurations, deployment contexts, and anticipated risks. Within this trend, the concept of a **prompt card** is gaining traction. A prompt card is a concise document that captures the purpose, structure, constraints, and ethical considerations behind a specific prompt. It enables engineers, designers, and reviewers to efficiently assess the intent and impact of a prompt within its context.

Major international organizations, such as the OECD, UNESCO, and ISO, have established AI ethics frameworks that emphasize explainable systems, anti-discrimination safeguards, and the preservation of human autonomy. Prompt engineering practices must now align with these broader commitments, especially in the public sector or sensitive domains.

12.12 Shared Responsibility in Practice

Though not policymakers, prompt engineers create systems that govern rights, accessibility, and trust. Each prompt functions as a de facto regulation that sets interaction norms, determines response limits, and shapes user expectations regarding validity, appropriateness, and truth. Consequently, prompt engineers must adopt a professional approach rooted in ethical frameworks and situational awareness. Engineers must articulate not only a prompt's function but also its design rationale, the affected stakeholders, and potential failure scenarios. They must disclose uncertainty, avoid deceptive framing, and collaborate with legal, policy, and domain experts to validate their work.

A well-crafted prompt is not merely one that "works": it serves user dignity, clarity, and accountability. Achieving this standard does not require perfection, but does require humility, a willingness to document decisions, acknowledge limitations, and revise designs in response to evidence, critique, or potential harm.

12.13 Reframing Prompting As Public Infrastructure

Prompting is no longer a fringe practice. It has become part of the underlying infrastructure of how societies communicate, organize knowledge, and make decisions. By designing these interactions, prompt engineers directly influence systems that regulate education, employment, and access to services.

This realization requires a shift in both professional identity and public framing. Prompt engineering is not just a creative act. It is now a form of civic design. It demands the same degree of care, transparency, and responsibility we expect from those who design public buildings, health systems, or financial regulations.

Future sections will offer specific professional guidelines, including a proposed code of practice. However, the premise begins here: prompt engineering now operates within the domain of public ethics. As such, it requires rigorous standards, institutional support, and a deep commitment to human-centered design.

12.14 Professionalization and the Prompt Engineer's Code

Prompt engineering has evolved into a role with substantial societal impact. As explored in previous chapters, prompts increasingly shape public services, educational experiences, healthcare communication, legal access, and workplace tools.

This widespread influence raises an essential question: *What does it mean to practice prompt engineering professionally?*

To answer this, we must distinguish between a technical skill and a professional discipline. Skills can be self-taught, task-specific, and evaluated by immediate performance. Professions, by contrast, carry a collective responsibility. Public trust, shared values, ethical standards, and an enduring dedication to societal benefit shape their work. Professions arise when a community recognizes that its work has social consequences and decides to steward that responsibility through principled standards and collaborative governance.

Prompt engineering now meets these criteria. The discipline influences how people access knowledge, receive guidance, make decisions, and form beliefs. Its consequences are not hypothetical; they are immediate, cumulative, and structural. Therefore, the field must advance not only in technical depth but also in ethical clarity.

12.15 From Competence to Care

The shift from skill to profession begins with a transformation in mindset. Prompt engineers must move beyond the question, *"What works best?"* and begin asking, *"What serves users with fairness, clarity, and dignity?"*

Practitioners must acknowledge that each prompt embeds assumptions about user identities, priorities, and anticipated behaviors. Some prompts may inadvertently exclude, confuse, or mislead. Others may marginalize alternative voices or reinforce dominant norms without transparency.

Professional practice requires conscious design and ethical *reflection*. The prompt engineer must explain the rationale behind a prompt, consider the diverse users it may affect, and evaluate its long-term implications when deployed at scale. This approach prioritizes diligence, responsibility, and iterative improvement over unattainable perfection.

This transition also calls for collective articulation. Like law, medicine, and architecture before it, prompt engineering needs ethical frameworks to safeguard both practitioners and the public. These should function not as inflexible rules, but as guiding principles, a professional compass that fosters responsible innovation and collective knowledge-building in this evolving field.

12.16 The Role of a Code of Practice

Ethical codes serve multiple purposes. They help communities align on values, navigate ambiguity, protect affected users, and uphold public confidence. In professions where practitioners make decisions with asymmetric knowledge or power, such as those of doctors, journalists, or data scientists, ethical codes help ensure that work serves the interests of people beyond the immediate client or task.

Prompt engineering fits within this model. The engineer may write a prompt alone, but that prompt enters a system, affects users, and may persist across thousands of interactions. The engineer may not be present to explain the prompt's intent or correct its failures. The code becomes the tether: a set of principles that connects design decisions to broader public accountability. A brief example later in this chapter demonstrates how a sample prompt card records rationale, assumptions, and safeguards without requiring visual diagrams.

Below, we offer a draft **code of practice**, not as a final prescription, but as a living framework for discussion, adoption, and evolution within the professional community.

12.17 A Draft Code of Practice for Prompt Engineers

Prompt engineers affirm that their work directly influences how people access information, experience care, and participate in digital systems. In this spirit, we propose the following commitments:

1. **Design with human dignity as the foundation.**

   Prompts must respect the tone, emotional needs, and identity of users. Especially in sensitive contexts, prompt engineers should prioritize empathy and respectful engagement.

2. **Prioritize understanding over performance.**

   A successful prompt not only generates output but also facilitates effective learning. It supports user comprehension, encourages critical thinking, and enables clarity, particularly for diverse audiences and those with varying access needs.

3. **Make assumptions visible.**

   Every prompt reflects design choices, including tone, audience, cultural references, and scenario. Prompt engineers must surface and document these assumptions to assess their inclusivity and potential impact.

4. **Ensure traceability and accountability.**

   Organizations must properly document, version, and maintain audit trails for system-embedded prompts, including their intended purpose, revision history, and identified risks, particularly for public or critical applications.

5. **The field must collectively embrace this stewardship responsibility.**

   Prompt engineers do not work in isolation. They collaborate with domain experts, community stakeholders, legal teams, and system designers to ensure the prompt serves institutional and societal goals.

6. **Practice professional humility.**

   Prompt engineers must recognize the limits of model capability and resist overstating system competence. They should disclose uncertainty, avoid speculative framing, and acknowledge knowledge gaps.

7. **Reject manipulation and harm.**

   Prompting should never be used to deceive, suppress dissent, or exploit vulnerable users. Engineers must resist pressures to optimize for engagement or compliance at the expense of ethical treatment.

8. **Support inclusion and shared learning.**

   The profession should facilitate accessibility by disseminating knowledge transparently, removing unnecessary gatekeeping, and amplifying marginalized perspectives. This ensures both practitioner diversity and equitable system design.

9. **Pause before deployment.**

   Before embedding prompts in systems, engineers should consider their effects at scale. Will this instruction promote trust, fairness, and sustainability? If not, it must be revised or withheld.

10. **Treat the code itself as a living commitment.**

    Ethical practice evolves. Prompt engineers must continually revisit, debate, and adapt their standards as the field evolves, new risks emerge, and global contexts change.

12.17.1 Becoming a Professional in Practice

Professionalism does not require certification. It requires care, judgment, and reflection. Prompt engineers already working in education, healthcare, civic tech, or cultural tools are shaping systems that mediate how people learn, seek support, and form trust. Whether or not they carry the title, they are part of a professional practice.

What matters now is how that practice is sustained: through shared education, collaborative review, and a commitment to the public good. Just as other disciplines have built institutions to govern their work, prompt engineering must do the same, ensuring that it remains a human-centered, ethical, and inclusive field in a rapidly evolving digital world.

12.18 Competency Map

The maturation of prompt engineering as a professional field requires not only conceptual clarity but also a structured developmental pathway. Readers frequently ask not only what prompt engineering entails but also how their competence can progress over time. To address this, a competency map provides a staged framework for learning and practice, guiding individuals from initial literacy to advanced systemic design.

Foundational literacy constitutes the entry point. At this level, learners become familiar with the mechanics of interacting with language models. They acquire the ability to distinguish vague from precise instructions, recognize how phrasing affects outputs, and identify basic risks such as bias or hallucination. The emphasis is not on technical depth but on building prompt awareness: understanding prompts as design inputs rather than casual queries.

The second stage, precision and architecture, introduces a deliberate structure. Here, learners practice constructing prompts that balance scope, constraints, and audience considerations. They learn to scaffold tasks, layer instructions, and apply reusable templates. At this stage, the practitioner moves beyond intuition toward repeatable design, appreciating the relationship between prompt form and communicative function.

The third stage focuses on CASTROFF mastery. Competence at this level entails fluent application of the CASTROFF framework as a comprehensive design discipline. Practitioners diagnose weak prompts across its dimensions, apply constraints systematically, and use CASTROFF not merely as a checklist but as an organizing principle for both analysis and creation.

The fourth stage, inclusive and safe practice, extends beyond technical proficiency to ethical responsibility. Competence here requires integrating inclusivity checks, red-flag lexicons, refusal scaffolds, and uncertainty statements into routine work. Practitioners also engage with adversarial testing, safety minimums, and debugging protocols to ensure outputs remain aligned with professional standards, especially in sensitive domains.

Finally, system-level design represents the most advanced stage. At this level, prompt engineers engage with prompts not as isolated artifacts but as embedded components of larger systems. Competence entails maintaining prompt registers, implementing version control and change protocols, and working within governance frameworks. This stage requires collaboration across disciplines and sensitivity to the institutional, legal, and cultural implications of deployed prompts.

The competency map is not a rigid hierarchy but a developmental trajectory. Practitioners may move unevenly through its stages, depending on context and role. An educator may emphasize inclusive practice before CASTROFF mastery, while a systems designer may leap quickly into governance structures. Nevertheless, the staged model offers clarity, progressing from literacy to precision, from framework mastery to safety, and ultimately to systemic integration.

By articulating this map, the field of prompt engineering signals its professional maturity. Competence is no longer measured solely by the ability to elicit functional responses from a model but by the capacity to align design intent with ethical responsibility, institutional governance, and long-term social trust.

Table 12-1. *Competency Map for Prompt Engineering*

Stage	Core Skills	Markers of Mastery
Foundational Literacy	Distinguish vague vs. precise prompts; recognize risks of bias/hallucination.	Produces consistent basic outputs; understands prompts as design inputs.
Precision and Architecture	Scaffold and layer tasks; balance scope, constraints, and audience.	Constructs structured prompts; adapts templates across contexts.
CASTROFF Mastery	Apply CASTROFF dimensions systematically for diagnosis and creation.	Uses CASTROFF as a design framework; identifies weak prompts across dimensions.
Inclusive and Safe Practice	Integrate inclusivity checks, refusal scaffolds, and safety protocols.	Anticipates harm; embeds safety measures; conducts adversarial and boundary testing.
System-Level Design	Maintain registers, apply versioning, and govern prompts institutionally.	Design prompts as infrastructural artifacts ensure accountability and traceability.

12.18.1 Role-Based Pathways of Competence

While the competency map provides a staged framework, practitioners often approach these stages from different professional vantage points. The following pathways illustrate how distinct roles may emphasize particular stages while still progressing across the broader spectrum of competence:

- **Practitioner pathway**: Builds steadily from foundational literacy through precision and architecture, with an emphasis on CASTROFF mastery for daily application.
- **Educator pathway**: Prioritizes inclusive and safe practice early, ensuring teaching materials and classroom prompts model ethical and accessible design.
- **Team lead pathway**: Moves quickly toward system-level design, focusing on documentation, governance, and collaborative protocols to manage prompt registers.
- **Policy and compliance pathway**: Intersects at inclusive and safe practice and system-level design, requiring fluency in regulatory standards and risk frameworks to guide organizational accountability.
- **Across sectors, similar competencies manifest differently**: healthcare emphasizes safety scaffolds; education foregrounds inclusive language, and civic services prioritize transparency and escalation.

These trajectories demonstrate that competence is not a linear process. Each role adapts the shared framework to its own responsibilities, while all converge on the professional goal of aligning prompt design with human dignity, institutional trust, and systemic accountability. A short example later in this chapter illustrates how an educator moves from basic literacy toward inclusive practice by progressively applying CASTROFF constraints in real classroom prompts.

12.19 Outlook

Looking forward, prompt engineering will expand beyond text into multimodal environments where speech, vision, and embodied interaction converge. It will also become integral to agentic workflows in which language models act as components within larger systems rather than as stand-alone tools. These developments will heighten the importance of governance standards, requiring prompt engineers to engage not only with technical design but also with institutional accountability and public trust. The discipline's future lies in striking a balance between advancing expressive possibilities and safeguarding human dignity and systemic reliability.

12.20 Recap

Chapter 12 marked the transition of prompt engineering from a technical curiosity to a recognized professional discipline. As language models become embedded in critical systems, shaping how people learn, access services, and make decisions, the practice of prompting must evolve beyond intuition or experimentation. It now demands shared standards, ethical reflection, and professional responsibility.

Section 12.1 outlined how prompting has evolved from informal trial-and-error to a structured design approach. What was once a flexible skill has matured into an interdisciplinary role with responsibilities across education, healthcare, law, and governance. Prompt engineers are no longer writing clever inputs for isolated outputs; they are shaping how systems engage with users at scale. This shift requires literacies in cognition, communication, ethics, and systems thinking.

Section 12.2 expanded the view of prompting from interaction to infrastructure. Prompts are increasingly embedded in tools, dashboards, civic services, and workflows, often invisible to the user but central to their experience. The prompt engineer's role now intersects with interface design, domain knowledge, and institutional accountability. Prompting is no longer about model cleverness; it is about human outcomes.

Section 12.3 examined the ethical and regulatory dimensions of prompting. Poorly constructed or hidden prompts can produce representational harm, misleading outputs, or inaccessible experiences. This section introduced emerging governance mechanisms, including the EU AI Act, prompt cards, and audit protocols. It argued that prompt engineers must document assumptions, design with care, and collaborate across disciplines.

Section 12.4 introduced a proposed *code of practice* for prompt engineers. This section made the case for professionalization, not through certification alone, but through values: transparency, traceability, user dignity, and collaborative responsibility. Prompting is no longer simply a technique; it has evolved into a civic practice that shapes how language models influence the public.

Section 12.5 presents a competency map that illustrates developmental stages, ranging from foundational literacy to system-level design, along with role-based pathways for practitioners, educators, team leads, and compliance specialists. It concluded with an outlook on multimodal prompting, agentic workflows, and governance standards, emphasizing that the profession's future lies in balancing innovation with responsibility.

Together, these sections establish prompt engineering as a field defined not only by its technical skill but also by its ethical stance, social impact, and commitment to inclusive, human-centered systems. Prompt engineers are no longer just system operators; they are stewards of communication in an AI-shaped world.

12.21 Further Reading

For readers seeking primary standards and authoritative references that inform the future of prompt engineering, the following sources provide valuable starting points:

- **European Union AI Act (2024)**: A comprehensive regulatory framework that classifies high-risk AI systems and requires human oversight, documentation, and accountability.
- **OECD AI Principles (2019)**: A set of international standards that emphasize human-centered values, transparency, and accountability in AI design.
- **UNESCO, Recommendation on the Ethics of Artificial Intelligence (2021)**: A global framework that emphasizes equity, inclusion, and human rights in AI deployment.
- **ISO/IEC JTC 1/SC 42, Artificial Intelligence Standards Portfolio**: Technical and governance standards for AI lifecycle management, risk, and transparency.
- **Mitchell, M. (2019). Artificial Intelligence: A Guide for Thinking Humans**: An accessible yet rigorous treatment of AI systems, their limitations, and social consequences.

12.22 Self-Assessment Quiz

1. Why is prompt engineering increasingly considered a professional discipline rather than just a technical skill?

    a) Because AI companies require formal certification for prompt engineers

    b) Because prompting now involves coding and system design

    c) Because prompt decisions carry public consequences and shape real-world outcomes

    d) Because the profession is officially recognized by international law

2. What is a significant ethical risk of hiding prompts within deployed systems?

    a) The prompts become outdated due to model drift.

    b) Developers may forget how to replicate results.

    c) The outputs cannot be explained, challenged, or audited by affected users.

    d) Prompts written in English disadvantage non-native speakers.

3. What is the core difference between a profession and a technical skill, as described in this chapter?

    a) Professions focus on individual expertise, while technical skills focus on teamwork.

    b) Professions regulate tools, while technical skills use them freely.

   c) Professions involve societal responsibility and shared ethical commitments.

   d) Professions are defined by university degrees and official credentials.

4. What purpose do "prompt cards" serve in professional prompt engineering practice?

   a) They provide automated summaries of output accuracy.

   b) They serve as UI templates for designing chatbot interfaces.

   c) They document the rationale, structure, and ethical assumptions behind a prompt.

   d) They instruct users on how to write prompts using natural language.

5. According to the proposed Prompt Engineer's Code, which of the following is a key professional commitment?

   a) To minimize user engagement time by streamlining output

   b) To design prompts that increase institutional control

   c) To deploy prompts immediately after performance testing

   d) To reflect on assumptions, disclose limitations, and serve user dignity

APPENDIX A

The CASTROFF Prompt Engineering Framework

Prompt engineering requires precision, clarity, and ethical mindfulness. The CASTROFF framework offers a structured approach to consistently design effective prompts for large language models.

Framework Components

- **C – Constraints**: Explicit limits on length, scope, or format.
- **A – Audience**: Intended reader or user context.
- **S – Structure**: Indicate the desired output format or layout (e.g., table, bullet list, code block).
- **T – Tone**: Specify the tone or voice for the response (e.g., formal, friendly, persuasive).
- **R – Role**: Identity or perspective assigned to AI.

H. Tavakoli, *Prompt Engineering for Everyone*,
https://doi.org/10.1007/979-8-8688-2338-1

- **O – Output format**: Set boundaries such as token or word limits, stylistic requirements, or exclusions.
- **F – Focus**: Thematic priority of response.
- **F – Function**: Communicative purpose or job of the outputs.

Professional Checklist

Before submitting a prompt, verify:

- Is the core task clearly and precisely defined?
- Have you assigned an explicit role for the model?
- Is the output format controlled and specified?
- Is the desired tone communicated?
- Have you identified the target audience?
- Are output limits clearly set?
- Have you included examples where appropriate?
- Have you accounted for risks and included safety filters?
- Is the task domain-specific enough to reduce ambiguity?
- Have vague instructions or filler phrases been removed?
- Is the prompt token-efficient?
- Are format markers used for machine readability?

- Have you chosen zero-shot or few-shot prompting intentionally?
- Is the prompt reproducible or ready for refinement?

Prompt Template

Fill in the blanks to structure your prompt professionally:

> You are a [ROLE].
>
> Your task is to [TASK GOAL] for a [TARGET AUDIENCE].
>
> Write in a [TONE] tone and use the [FORMAT] format.
>
> Keep it under [LENGTH] tokens/words and include [REQUIREMENTS].
>
> Avoid [EXCLUSIONS].
>
> Respond only with the output (no extra explanation).
>
> [Insert one to two examples if needed for a few-shot pattern.]

Cost-Saving Heuristics

- Use zero-shot prompting for simple, well-defined tasks.
- Avoid unnecessary adjectives, politeness formulas, or elaborate introductions.

- Combine instructions and formatting in one sentence where possible.
- Use token-efficient formats such as key/value pairs instead of paragraphs.
- Limit explanations to what is strictly necessary.

APPENDIX B

Advanced Prompt Engineering Checklist

This checklist is designed for professionals aiming to craft prompts that are precise, controlled, and ethically sound. It supports the creation of prompts that reliably generate high-quality outputs while respecting context, audience, and fairness.

1. Pre-prompting Preparation

- **Define the purpose**: What specific problem or task is the prompt addressing?
- **Identify the context**: Where and how will the generated output be used?
- **Understand your audience**: Who will read or act on the output?
- **Choose the appropriate model**: Consider the capabilities and limitations of the language model in use.

H. Tavakoli, *Prompt Engineering for Everyone*,
https://doi.org/10.1007/979-8-8688-2338-1

2. Prompt Composition Strategies

A. Role-Specialist Prompting

- Assign a clear, domain-specific role early in the prompt (e.g., "You are a data privacy officer").
- Avoid stacking multiple roles; maintain one focused identity.
- Use the role to guide style, tone, and content depth.

B. Tone Control

- Specify tone keywords such as "formal," "friendly," "instructional," or "persuasive."
- Combine tone with role and audience for a consistent voice.
- Example prompt structure:

  "Write in a [tone] tone for a [role] speaking to a [type of audience]."

C. Output Format Prompting

- Specify the required output format: bullet points, tables, letters, JSON, markdown, etc.
- Use dummy headings to outline structure without examples, e.g.:

```
## Strengths
## Weaknesses
```

- Be explicit with instructions such as "Do not include preamble" or "Return only the table."
- Use formatting strategically to control token usage and focus.

D. Framework Prompting

- Embed expert reasoning models such as SWOT, STAR, AIDA, 5 Whys, MECE, or RACI.
- Name frameworks explicitly and optionally explain components.
- Chain frameworks sequentially for deeper analysis.
- Reinforce with dummy headings or structured layouts.

E. Focus Prompting

- Limit scope clearly with phrases like "Only list ethical concerns" and "Exclude technical details."
- Use instructions such as "Avoid," "Limit to," or "Do not elaborate on."
- Set output boundaries: "No more than three points" and "Do not exceed five lines."
- Reinforce focus by restating the task and target areas.

3. Model-Aware Prompting Tricks

- Use trigger phrases such as "Let's think step by step" to encourage reasoning (Zero-CoT).
- Include dummy format blocks (e.g., JSON) to improve structure compliance.
- Place formatting instructions at the end of the prompt to strengthen memory.
- Break down complex tasks with numbered steps:

  1. List key issues.
  2. Summarize the top three.
  3. Provide a final assessment.

4. Efficiency and Ethics

- Avoid over-prompting; overly long prompts can confuse the model or waste tokens.
- Prioritize clarity and simplicity; remove ambiguous language.
- Consciously avoid wording that may trigger bias or hallucinations.
- Respect downstream applications by prompting with fairness and safety in mind.

Optional Add-Ons

- Use few-shot examples sparingly when an exact output style is needed.
- Include scaffolding instructions such as

  "Do not continue until all three sections are complete."
- Add self-reflection cues, for example:

  "Before answering, check for logical consistency."

References and Further Reading

1) Angwin, J., Larson, J., Mattu, S., & Kirchner, L. (2016). Machine Bias. ProPublica. https://www.propublica.org/article/machinebias-risk-assessments-in-criminal-sentencing

2) Buolamwini, J., & Gebru, T. (2018). Gender Shades: Intersectional Accuracy Disparities in Commercial Gender Classification. MIT Media Lab. https://www.media.mit.edu/publications/gender-shades-intersectional-accuracy-disparities-in-commercial-genderclassification

3) Dwork, C., Hardt, M., Pitassi, T., Reingold, O., & Zemel, R. (2012). Fairness Through Awareness. In Proceedings of the 3rd Innovations in Theoretical Computer Science Conference (ITCS'12), ACM, pp. 214–226. http://doi.acm.org/10.1145/2090236.2090255

4) Dressel, J., & Farid, H. (2018). The Accuracy, Fairness, and Limits of Predicting Recidivism. Science Advances, 4(1), eaao5580. https://www.science.org/doi/10.1126/sciadv.aao5580

H. Tavakoli, *Prompt Engineering for Everyone*,
https://doi.org/10.1007/979-8-8688-2338-1

5) Floridi, L., & Cowls, J. (2019). A Unified Framework of Five Principles for AI in Society. Harvard Data Science Review. https://doi.org/10.1162/99608f92.8cd550d1

6) Hardt, M., Price, E., & Srebro, N. (2016). Equality of Opportunity in Supervised Learning. NeurIPS 2016. https://arxiv.org/abs/1610.02413

7) Leben, D. (2025). AI Fairness: Designing Equal Opportunity Algorithms. MIT Press. https://mitpress.mit.edu/9780262552363/ai-fairness/

8) Chang, K., Xu, S., Wang, C., Luo, Y., Liu, X., Xiao, T., & Zhu, J. (2024). Efficient Prompting Methods for Large Language Models: A Survey. arXiv. https://arxiv.org/abs/2404.01077

9) Liu, P., Yuan, W., Fu, J., Jiang, Z., Hayashi, H., & Neubig, G. (2021). Pre-train, Prompt, and Predict: A Systematic Survey of Prompting Methods in NLP. arXiv. https://arxiv.org/abs/2107.13586

10) Liu, X., et al. (2021). P-Tuning v2: Prompt Tuning Can Be Comparable to Fine-Tuning Universally Across Scales and Tasks. arXiv. https://arxiv.org/abs/2110.07602

11) Mehrabi, N., Morstatter, F., Saxena, N., Lerman, K., & Galstyan, A. (2021). A Survey on Bias and Fairness in Machine Learning. ACM Computing Surveys, 54(6), 1–35. https://dl.acm.org/doi/10.1145/3457607

12) OECD. (2019). OECD AI Principles: Recommendation of the Council on Artificial Intelligence. https://www.oecd.org/going-digital/ai/principles/

13) Ouyang, L., et al. (2022). Training Language Models to Follow Instructions with Human Feedback. arXiv. https://arxiv.org/abs/2203.02155

14) Brown, T.B., Mann, B., Ryder, N., Subbiah, M., Kaplan, J., Dhariwal, P., . . . Amodei, D. (2020). Language Models are Few-Shot Learners. arXiv. https://arxiv.org/abs/2005.14165

15) Bohr, A., & Memarzadeh, M. (2023). Prompt Engineering as an Important Emerging Skill for Medical and Healthcare Professional Practice. Journal of Medical Internet Research, 25, e50638. https://www.jmir.org/2023/1/e5063

16) Zeit Online. (2024). Unleashing the Potential of Prompt Engineering for LLMs. ScienceDirect. https://www.sciencedirect.com/science/article/pii/S2666389925001084

17) ISO/IEC JTC 1/SC 42. (2021). Information Technology—Artificial Intelligence—Overview of Trustworthiness in AI (ISO/IEC TR 24020). https://www.iso.org/standard/77608.html

18) Zeit Online. (2025). AI Fairness Manifesto: Designing for Equal Opportunity. Time Magazine. https://time.com/6756770/ai-biasalgorithm-for-equality/

19) Zhou, Y., et al. (2022). Large Language Models Are Human-Level Prompt Engineers. arXiv. https://arxiv.org/abs/2211.01910

Afterword

This book has traced a fundamental shift in the role of prompt engineering: from a tactical technique for eliciting responses to a systemic responsibility in the design of human–AI interaction. What begins as linguistic instruction now shapes institutional processes, structuring how knowledge is retrieved, how choices are framed, and how humans experience the logic of machine-mediated systems.

This transformation carries consequences well beyond usability. As prompts become embedded in tools that influence clinical decision-making, educational access, bureaucratic procedures, and workplace communication, they acquire social and ethical force. A prompt does not merely generate an output; it defines a possibility space. In doing so, it shapes how options are perceived, how authority is communicated, and how fairness is enacted, or denied.

Across professions, individuals are increasingly called upon to design, interpret, or respond to AI-generated content. These individuals are not necessarily AI researchers or machine learning engineers. More often, they are educators, clinicians, policy analysts, service designers, researchers, and public servants, people whose primary commitments lie in equity, care, and clarity. This book has been written for them: to support principled reasoning in prompt design, critical evaluation, and responsible revision.

The human-centered perspective developed throughout this text insists on the importance of context, interpretability, and inclusive design. Prompting, in this light, is not a neutral act. It is a form of epistemic mediation, a way of shaping what counts as relevant, accurate, or acceptable knowledge. To treat prompting as a professional discipline is to recognize its influence on meaning, identity, and access.

H. Tavakoli, *Prompt Engineering for Everyone*,
https://doi.org/10.1007/979-8-8688-2338-1

Yet prompt engineering remains an emergent field. It lacks universally accepted standards, formal qualifications, and globally endorsed governance frameworks. What it does not lack, however, is consequence. In response to this growing consequence, a new expectation is taking shape: that prompting be treated not as a private art or opportunistic exploit, but as a transparent, reviewable, and justifiable process. This expectation does not require institutional gatekeeping. It requires intentionality, accountability, and care.

The professionalization of prompt engineering begins not with regulation, but with reflection. It starts with the recognition that prompting shapes how systems serve people and that this shaping must be guided by values we are willing to justify and uphold. In proposing transparency, inclusion, traceability, and fairness as foundations of prompt practice, this book does not claim final authority. Instead, it contributes to an ongoing effort: the collective development of standards worthy of the trust that language technologies now command.

As language models are deployed across domains of public significance, such as policy, education, health, and governance, the design of their instructions becomes inseparable from the ethical and institutional futures they help shape. Whether or not the term *"prompt engineer"* becomes a formalized role, the responsibilities associated with it are already real. They belong to anyone shaping how AI interacts with the world.

This book closes not with resolution, but with commitment: a commitment to help ensure that the systems we build remain accountable to the people they serve. The future of prompt engineering will not be defined solely by scale or innovation. It will be shaped by the integrity, humility, and human dignity with which its practitioners act.

A Call to Educators, Public Institutions, and Civil Society

The responsibility for human-centered prompting cannot rest solely with private developers. It must also be taken up by public institutions, civic organizations, and educators, those who shape how technologies are interpreted, implemented, and understood. This is especially urgent in multilingual societies and under-resourced contexts, where language barriers risk becoming structural barriers to participation in AI systems.

Educational institutions are well-positioned to treat prompting as a foundational literacy, one that complements writing, critical thinking, and ethical reasoning. Government agencies and public service providers must invest in inclusive prompt infrastructure, including shared templates, multilingual guidance, and accessible design practices tailored to healthcare, justice, education, and civic engagement. International bodies that aim to reduce digital inequality must support research and training in culturally responsive prompting, developed in collaboration with underrepresented communities.

Institutions of care, law, and learning must begin to treat prompts not as minor technical details but as operational elements of trust, clarity, and justice.

To delay such efforts is to widen the gap between those who shape AI systems and those who are shaped by them. The values of prompt engineering—clarity, care, inclusion, and foresight—must be enacted now, while the field is still forming and before systemic harms are further entrenched.

We must not wait for fluency in dominant global languages before designing AI that serves all communities. We must build systems that reflect the full dignity, diversity, and linguistic richness of the human world.

This work is offered in support of that vision.

Postscript

This book was written at a pivotal moment in the public life of artificial intelligence. During its development, the capabilities, reach, and risks of language models expanded rapidly, often unpredictably, outpacing regulation, pedagogy, and public understanding.

In such a dynamic environment, no single volume can claim to provide a comprehensive or definitive account. What this book offers instead is a foundation: a principled, accessible, and ethically grounded introduction to prompt engineering for those working outside the core of AI research and development. Its central argument, that prompt design is both a technical and moral act, remains relevant, even as the tools themselves evolve.

The reader who carries this work forward is not merely adopting techniques; they are also engaging with the ideas presented. They are helping shape the standards by which those techniques are evaluated. That future cannot be authored solely by developers. It depends on the participation of educators, clinicians, civil servants, researchers, artists, students, and advocates, anyone with a stake in how language is used to shape systems that serve human needs.

This book does not offer the final word. It is an invitation: to take responsibility for the language we give to machines and for the systems those words help create.

H. Tavakoli, *Prompt Engineering for Everyone*,
https://doi.org/10.1007/979-8-8688-2338-1

Index

A

B

H. Tavakoli, *Prompt Engineering for Everyone*,
https://doi.org/10.1007/979-8-8688-2338-1

C

D

E

F

G

H

I

J, K

L

M

N

O

P, Q

T

U

V, W, X, Y

Z